Helion & Company Limited
Unit 8 Amherst Business Centre
Budbrooke Road
Warwick
CV34 5WE
England
Tel. 01926 499 619
Email: info@helion.co.uk
Website: www.helion.co.uk
Twitter: @helionbooks
Visit our blog http://blog.helion.co.uk/

Designed and typeset by Farr out Publications, Wokingham, Berkshire
Cover design by Paul Hewitt, Battlefield Design (www.battlefi eld-design.co.uk)

ISBN 978-1-914377-04-4

British Library Cataloguing-in-Publication Data
A catalogue record for this book is available from the British Library

We always welcome receiving book proposals from prospective authors.

CONTENTS

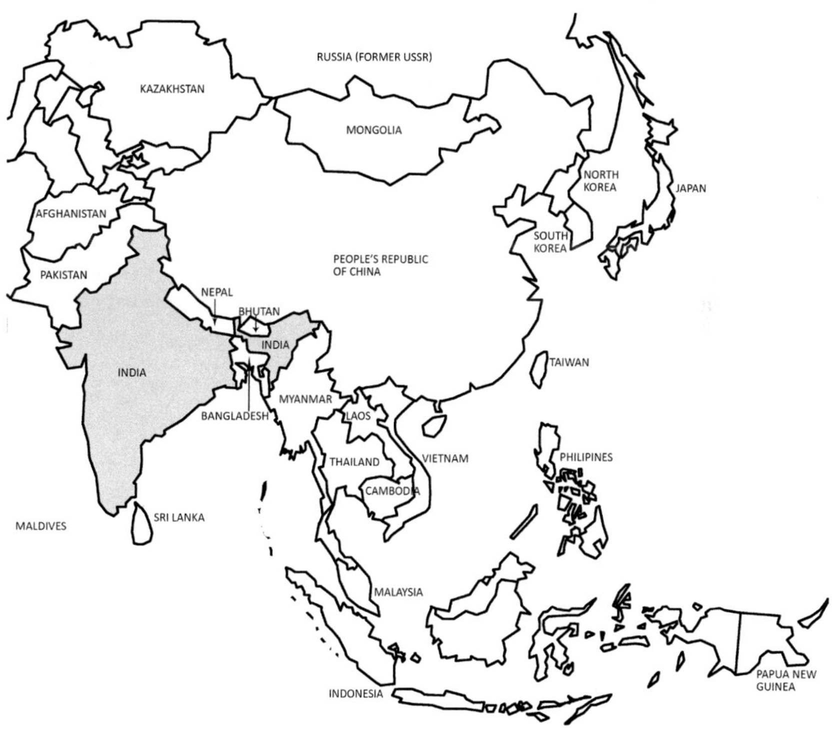

Note: In order to simplify the use of this book, all names, locations and geographic designations are as provided in *The Times World Atlas*, or other traditionally accepted major sources of reference, as of the time of described events.

LIST OF ABBREVIATIONS

AAD	Advanced Air Defence
ADCC	Air Defence Control Centre
ADGES	Air Defence Ground Environment System
AEC	Atomic Energy Commission
ARDE	Armament Research and Development Establishment
BADZ	Base Air Defence Zone
BMD	Ballistic Missile Defence
BARC	Bhabha Atomic Research Centre
BJP	Bharatiya Janata Party
CBRN	Chemical, Biological, Radiological, Nuclear
CEP	Circular Error Probable
CO	Commanding Officer
CRC	Control and Reporting Centre
DRDO	Defence Research and Development Organisation
IAF	Indian Air Force
INC	Indian National Congress
IRBM	Intermediate-Range Ballistic Missile
ICBM	Intercontinental Ballistic Missile
LAC	Line of Actual Control
LoC	Line of Control
MANPADS	Man Portable Air Defence System
MOP	Mobile Observation Post
MRBM	Medium-Range Ballistic Missile
NDMA	National Disaster Management Authority
NDRF	National Disaster Response Force
PIB	Press Information Bureau (Indian Government)
POK	Pakistan Occupied Kashmir
SAM	Surface to Air Missile
SLBM	Submarine-Launched Ballistic Missile
SRBM	Short-Range Ballistic Missile
TBRL	Terminal Ballistics Research Laboratory
TEL	Transporter-Erector-Launcher
TNW	Tactical Nuclear Weapon
UAV	Unmanned Aerial Vehicles

INTRODUCTION

Of the world's nuclear powers – eight declared (the United States, Russia, China, the United Kingdom, France, India, Pakistan and North Korea) and one undeclared (Israel) – India's path to nuclear weapons status and the current state of its arsenal have been the subject of many books and interpretations of its intention. Alone among nuclear powers, India has no history of nuclear proliferation and professed a commitment to global nuclear disarmament which lasted for decades. Now, as India moves towards the development and deployment of a full nuclear triad with its attendant command structures for three modes of delivery, some degree of understanding of India's path to nuclear weapons deployment and its nuclear doctrine are essential.

In this book, which at best is a brief overview, a short synopsis will be given of India's security policies and their evolution over the decades, along with the role that nuclear weapons played in that evolution. India's security challenges with Pakistan and China will be detailed, along with its conflicts with these countries as well as the contingencies it has made to deal with the challenges they pose.

India's first nuclear test in 1974 and its hesitant and decidedly slow move towards weaponisation in the 1980s will be discussed, with emphasis on the debates within India and the hesitancy towards weaponisation. India's weaponisation in the mid-1990s, along with the development of India's air-delivered and ballistic missile nuclear weapons, leading to the decision in 1998 to resume nuclear testing, will be analysed. Subsequently, India's deployment of a small nuclear deterrent during the 1999 Kargil War against Pakistan will also be discussed. The majority of the book deals with three core areas of the Indian nuclear arsenal: the current state of the arsenal in respect of warheads, fissile material and delivery systems; India's nuclear doctrine and strategy; and India's quest for a ballistic missile defence with its attendant implications. India's nuclear weapons saga and its current capabilities are little known and largely misunderstood in the West, but it is hoped that this brief narrative and analysis will help to place such matters in a better perspective.

India's Security Dilemma

Central to India's security is the fact that it is bordered by two countries with which it has serious border disputes. On the one hand, India has a volatile border with Pakistan and is wrestling with a Pakistani-sponsored insurgency in India's part of the disputed Kashmir region. With China, on the other hand, the border dispute is not nearly as serious, though there is mutual distrust between the two nations and a substantial Indian troop presence has been maintained along the Sino–Indian border since the two countries fought a short but vicious war in 1962. India has thus had to arrange its conventional and non-conventional forces to deal with perceived threats from both Pakistan and China. The bulk of India's forces are arrayed against Pakistan, but a substantial troop presence is meanwhile maintained against China. The Sino–Pakistan strategic partnership has created significant concern in India, since China has had few qualms about selling missile technology to Pakistan and has possibly even supplied Pakistan with reliable designs for atomic weapons. Indeed, it is the perceived threat of the Sino–Pakistan nexus that prompted India to continue with the development of its own nuclear weapons programme.

China is an established nuclear power, having first detonated a nuclear weapon in 1964. This provided the initial impetus to the Indian nuclear programme, and is still used as the primary justification for the 'Agni' Intermediate Range Ballistic Missile programme. Though India and China have not gone to war since 1962, and although India has dramatically improved its conventional military position against China – to the extent where India holds limited tactical superiority and virtually numerical parity in the Sino–Indian theatre – India remains suspicious of Chinese intentions and perceives itself to be at least under a peripheral threat from Chinese nuclear weapons.

With regards to Pakistan, the situation is somewhat different. Two major wars were fought, first in 1965 and then in 1971 where India scored a major victory over Pakistan's forces in what was then East Pakistan, but in neither war was India able to, nor did it really try to, decisively defeat Pakistan in the west. Nuclear weapons became a major part of the India–Pakistan equation during the winter of 1986/1987. India long believed, with some justification, that Pakistan had been fomenting terrorism and insurgency in India since the mid-1980s, and still is. This was particularly the case in Punjab, though there were some signs in the late 1980s of the emergence of the Kashmir insurgency. India believed that Pakistan sought the balkanisation of India, or at the very least, to create such internal turmoil in India so as to facilitate a Pakistani territorial grab in Kashmir. This was, and is still, used to justify the existence and continual modernisation of India's huge conventional military forces.

The Indian conventional forces have always outnumbered their Pakistani counterparts by a substantial margin. Moreover, India also began experimenting with new tactics involving the use of massed armour and mechanised infantry formations and, in 1986–87, conducted a series of military exercises that would facilitate the bisection of Pakistan. In response to this, Pakistan brought itself to the verge of nuclear weaponisation so as to deter India. This, in turn, sparked a resurrection of India's own nuclear weapons programme so as to prevent any nuclear blackmail on the part of Pakistan. The India–Pakistan nuclear race that started more than a decade ago has acquired a momentum all of its own, with neither country trusting the other sufficiently to give up its nuclear option. Although China was the principal reason for India initiating a nuclear weapons programme, Pakistan has now overtaken China in India's nuclear and conventional threat perceptions owing to the rise in cross-border terrorism and tension over Kashmir.

So where do nuclear weapons fit into India's threat perceptions? India views nuclear weapons as an ultimate deterrent against both conventional military threats and the fear of balkanisation, from domestic insurgency sponsored by foreign powers, as well as providing a deterrent against nuclear attack. In one way, nuclear weapons are very much a part of India's internal security apparatus. This does not mean that India would use a nuclear weapon against an insurgent group. Rather, it serves to warn foreign powers – namely Pakistan and China – that India could, and would, inflict unacceptable damage on them should they attempt to snatch parts of India.

The creation of defence policy for an independent India dates back to 1945, when the British colonial authorities in India began demobilising the 2.5 million-strong Indian Army of the Second World War. Most British and Indian military writers and analysts argued that a regional approach to Indian security should be adopted.[1] One proposal was for India to form the nucleus of a regional Commonwealth force. These proposals led to a reassessment of the required size and strength of the Indian armed forces.

The Indian Army was to be reduced from a size of 2.5 million to about 10 infantry divisions, an armoured division, an airborne division, 50 frontier-defence battalions and 63 internal security battalions. The Indian Navy was to be built around a nucleus of three cruisers, while the Indian Air Force was to be doubled to a strength of 20 squadrons. However, this plan came to nought because of the turbulence in the Indian subcontinent. Consequently, the armed forces were compelled to commit themselves to the difficult task of protecting refugees and establishing relief camps. Moreover, the partition of British India into India and Pakistan had a direct impact on the Indian armed forces, which were divided between India and Pakistan by a ratio of two to one, effectively ending any attempts to initiate the reorganisation plan mentioned above. India's share amounted to 280,000 men in an army of four infantry divisions with some artillery and armour; a navy of 11,000 personnel with four sloops, two frigates and one corvette; and an air force of seven fighter squadrons and a single transport squadron.[2]

The new Indian government, led by Jawaharlal Nehru, was unfamiliar with defence issues. Nehru, along with many of the leadership of the Indian National Congress (INC), had spent much of the Second World War in prison due to his support for civil disobedience and demands for the British to leave India, and played no role in formulating defence policy. As a result, Indian defence policy in the post-war period was formulated by an inexperienced government.

After independence, Prime Minister Nehru argued that India's peace and security could not be ensured through alliance systems and military expansion.[3] Instead, Nehru proposed to build India's defence and foreign policies around the five principles of 'Panchseel'. These principles of peaceful coexistence with communist states were first embodied in the Sino–Indian Treaty of 1954, as Nehru's government was keen to establish good relations with China.

India's policy of non-alignment had serious consequences for the Indian armed forces. The Indian military, unlike their Pakistani counterparts, were unable to obtain significant United States military aid because of India's friendship with communist countries. A rather more serious effect of the 'Panchsheel' policy was that India's border with China was not adequately defended. While there was the purchase of some new equipment, there was no expansion of the Indian armed forces, and defence budgets for essential maintenance and war wastage reserves were often savagely cut.

The Indian government had requested that the former British Viceroy of India – Lord Louis Mountbatten – become Governor-General immediately after independence to lend his military expertise so that effective convoys for refugees from the conflict in Kashmir could be organised. Even this task was beyond the existing expertise of the new Indian government.

Due to India's war with Pakistan over Kashmir between 1947 and 1949, almost all of India's combat units and military infrastructure were designed to meet the Pakistani threat. India's defence plans assumed that Pakistan would initiate action in Kashmir, leading to an Indian counterattack along other fronts. The Indian Army was to concentrate its full strength against Pakistan, to defeat the Pakistani Army and, if necessary, to occupy or destroy the Pakistani cities of Lahore and Sialkot.[4]

The possible threat from the People's Republic of China was deliberately downplayed by the Indian government, which was attempting to forge a friendly relationship with the then quasi-pariah communist regime. In fact, army units in eastern India were either deployed for counterinsurgency duties in the troubled North-Eastern States or for the defence of Calcutta. The defence of the China border was entrusted to the lightly armed personnel of the Central Reserve Police Force. Artillery and logistical support was non-existent, and communications – both by radio and by road – were precarious at best. There were a number of Indian Air Force bases located in the vicinity of the north-eastern border with China, but these lacked infrastructure and few combat aircraft were based there.

The Sino–Indian border war of 1962 came as a shock to the Indian government, though not really to the Indian military. Along the north-east frontier, Indian forces were defeated after a number of ill-conceived command decisions, while the induction of poorly equipped reinforcements, none of whom were properly acclimatised, did nothing to halt the Chinese advance. The government had earlier ignored demands by the Army for the creation of dedicated mountain divisions and the building of suitable infrastructure to allow for a successful defence of the border with China.[5]

The exception to this general trend of humiliating defeats was the situation in Ladakh. Here, an army brigade had been in position since 1959 and had devised a reasonably effective layered defence plan. Though some tactical withdrawals to secondary defensive positions occurred, Indian forces remained cohesive and effective combat units in the Ladakh sector.

The aftermath of the humiliating defeat suffered by India in this short but sharp border war with China saw a major reassessment of India's defence policy. The Indian government recognised that it had to contend with both Pakistani and Chinese military threats, and as such, troops, equipment, defence plans and an effective defence policy had to be provided to deal adequately with the new strategic environment. This has remained the cornerstone of Indian defence policy ever since, the dual threats from Pakistan and China constantly being cited by the Indian leadership, though China is perceived to be a much less likely adversary. Moreover, one of the reasons for the massive military reorganisation and modernisation programme that ensued was to ensure that the Indian government and its politicians would not have to face the fury of the Indian public, which was of course also the electorate, if another military defeat occurred.

The post-war period saw a dramatic increase in size and capability of the Indian armed forces. Although the expansion was comprehensive, it was still essentially defensive. The thrust of Indian defence policy was evolved around what R.G.C. Thomas calls the 'minimalist' perspective of being able to fight one 'full' war and one 'half' war.[6] In military terms, this meant that India was capable of fighting a war on land, sea and in the air against Pakistan, while being able to fight a defensive border war with China.

Unlike the situation in the 1980s and 1990s, though perspective has once again altered in the post-2010 period, China was seen as ultimately the most significant threat throughout the 1960s and 1970s. Although the 1965 and 1971 wars with Pakistan proved that the Pakistani forces were still a substantial adversary, India's leadership viewed China's increasing power with growing concern. To this end, despite the substantial growth in India's naval capability and a widening gap in Indian air superiority, India has still confined itself to being able to fight a full-scale offensive war with Pakistan, while deploying sufficient forces to conduct only a defensive war against China.

Even the growth in India's nuclear and missile capability, which has now reached the stage where India has a considerable ballistic missile potential and a large stock of fissile material, has to be seen in light of the fact that there has been no perceptible change in India's basic strategy regarding China and Pakistan. Nuclear weapons are an extension of this one-and-a-half wars capability. No meaningful power projection doctrine has yet emerged in India.

India has moved on from this position to one where it has the capacity to fight one full and three half-wars. This means that India would be able to conduct a full war with Pakistan and a defensive border war with China, combined with a latent nuclear weapons capability and the ability to intervene in neighbouring Indian Ocean islands.[7] This may be considered to be India's current strategic position. India has deployed IRBMs (Intermediate-Range Ballistic Missiles) and is openly moving slowly towards full-range ICBMs (Intercontinental Ballistic Missiles). However, there seems to be no evidence of India shifting its basic threat perceptions in the near future. Therefore, Pakistan, and to a lesser extent China, remain, in the view of the Indian government, the two major military threats faced by India; to this end, both substantial conventional forces and a significant nuclear potential are seen to be vital to preserving India's ability to deter an attack from either country.

The Beginning

India's nuclear programme can trace its ancestry to the creation of the Institute for Fundamental Research (IFR) in 1945. It was led by a brilliant and enthusiastic Cambridge-educated physicist, Dr Homi J. Bhabha, who was to have an enormous influence on nuclear research for both military and civilian purposes.[8]

The IFR's interest in nuclear energy found a receptive audience in Prime Minister Jawaharlal Nehru, who saw nuclear energy as a source of power capable of fuelling India's economic development and industrial progress. The peaceful uses of nuclear energy was emphasised by both Nehru and Bhabha. Dr Bhabha declared that 'When nuclear energy has been successfully applied for power production in, say, a couple of decades from now, India will not have to look abroad for its experts, but will find them ready at home.'[9]

Self-reliance was the cornerstone of India's nuclear programme, and Bhabha and his colleagues were tasked with developing a broad-based, predominantly self-reliant nuclear sector which would be the spearhead of technological development in other sectors of the Indian economy. One year after independence, the Indian Atomic Energy Commission (AEC) was set up with a mandate to develop a comprehensive nuclear programme for civilian purposes.

Headed by Bhabha, the AEC, was to become the linchpin of the current Indian nuclear infrastructure. The Indian government gave the AEC clear guidelines that self-reliance was to be the ultimate objective: 'India should be able to produce all the basic materials required for the utilization of atomic energy ... The special requirements of atomic energy, the newness of the field, the strategic nature of its activities and its international and political significance have to be borne in mind.'[10]

By emphasising the peaceful aspects of the nuclear programme, Indian scientists were able to learn a great deal from Western countries, especially Canada, France and the United States. In 1951, the French signed a nuclear cooperation agreement that provided for the training of Indian nuclear scientists in France. The Indian scientists were also instructed in the process of extracting plutonium from spent nuclear fuel – a process critical to the manufacture of atomic bombs with plutonium cores.

The United States and Canada provided a great deal of training and equipment for Indian scientists, some of this assistance being provided under the 'Atoms for Peace' scheme of the Eisenhower administration.[11] The United States began cooperating with India for the construction of two nuclear power reactors at Tarapur, while the Canadians were responsible for the supply of fuel to the CIRUS reactor which was built in India with extensive Canadian assistance. These reactors were critical to the early stages of India's nuclear research and weapons programmes.

India had the distinction of commissioning Asia's first nuclear reactor outside the then Soviet Union. The Apsara research reactor was indigenously built in India and started operating in 1956.[12] Located at the Bhabha Atomic Research Centre (BARC), the reactor was a light-water/medium-enriched uranium type which used uranium enriched to 20–50 percent as a fuel source. As India lacked any sort of uranium enrichment facility at that time, the United Kingdom supplied the fuel under a 1955 contract.[13]

The Apsara was followed by the CIRUS 40MW (megawatt) research reactor, which paved the way for nuclear cooperation with both the United States and Canada. This reactor was constructed with considerable assistance from Canada, although the major part of the technical effort involved in the construction of CIRUS came from India. As the reactor was a heavy-water/natural uranium type, India initially needed external assistance for both the uranium fuel and the heavy-water coolant.[14]

Under a 1956 contract, Ottawa provided the initial requirement of fuel, but, since Canada had not yet fully developed its own heavy-water industry, the United States – under the auspices of President Eisenhower's 'Atoms for Peace' policy – signed an agreement for the sale of 18.9 metric tonnes of heavy-water on 16 March 1956.[15] In order to ensure that their assistance was not used for the development of nuclear weapons, both the United States and Canada obtained from India assurances that the heavy-water, the reactor and the products obtained from it were to be used only for peaceful purposes.[16] There was, however, one very serious flaw: none of the agreements barred, either explicitly or implicitly, the use of the material in peaceful nuclear explosions (PNEs).[17] Dr Bhabha consciously sought to avoid any agreements that would compromise, restrict or limit India's complete control over the reactor and its output.[18]

Prime Minister Indira Gandhi with aides visiting the site of the Pokhran-1 test. (Press Information Bureau, PIB)

The Chinese Invasion Shock

The Chinese invasion of 1962 was sudden and totally unexpected by India's government. The Indian armed forces, suffering from years of neglect and from a fundamentally flawed forward deployment plan, was defeated in a series of bruising encounters in India's north-east and was under heavy pressure in the northernmost regions of the country. Nehru's government then began a major review of India's defence requirements. This marked the beginning of India's quest for a nuclear weapons capability, though it would be more than a decade before India's first nuclear test.

Almost exactly two years later, on 16 October 1964, China tested a 20-kiloton atomic bomb. This sent shockwaves through the Indian establishment, and calls were made for the deployment of a nuclear arsenal. The panic generated forced Dr Bhabha to give public assurance that India could produce nuclear weapons within 18 months of receiving political clearance.[19]

Nehru's successor, Lal Bahadur Shastri, came under increasing pressure from Bhabha and the rest of the nuclear establishment to carry out an underground test in January 1965. Shastri successfully resisted such pressure, but, shortly after the Chinese test in 1964, he is believed to have given the go-ahead for a project to reduce the lead time for the production of nuclear weapons from 18 months to just six.[20]

India's technical capacity had increased to a point where, by June 1965, enough plutonium for an atomic bomb would have been produced. Moreover, India was not neglecting the production of materials needed for thermonuclear weapons. By 1970, India would have the technical capacity to manufacture one bomb per week.[21] This increasing capability was largely due to the early assistance of Canada and the United States and the capacity of the CIRUS reactor. Moreover, by 1966, India had acquired the ability to reprocess plutonium from the CIRUS reactor in the form of a plutonium reprocessing plant, constructed with assistance from the United States, which was then capable of reprocessing 30 metric tons of spent fuel per year.

The Indian military was rather more cautious about nuclear weapons, being fully occupied with a massive expansion and modernisation of conventional military forces. However, some middle-ranking officers made fairly strident calls for tactical nuclear

Prime Minister Gandhi being assisted by scientists at the test site. (PIB)

weapons to be deployed to thwart any Chinese attack on India's north-eastern frontiers.[22]

An extremely articulate and well-reasoned argument in favour of India developing nuclear weapons was made by Colonel R.D. Palsokar in 1966. Assuming a Chinese attack on India's north-eastern regions, Palsokar created a scenario where low-yield nuclear weapons were used by China to secure a tactical breakthrough. In this scenario, Palsokar argued that India would be unable to halt a Chinese advance and that huge tracts of territory would be lost. To deal with this threat, he advocated the deployment of a limited number of nuclear weapons. Palsokar, along with many other retired military officers, argued that a small minimum deterrent would easily be within India's technical and economic reach.

Quoting Major General D. Som Dutt, Palsokar argued that a deterrent force of 50 plutonium bombs, accrued at the rate of five per year and without an expensive missile-based delivery system, would cost about $500 million over a 10-year period. Dr Bhabha argued that a stockpile of 50 two-megaton thermonuclear bombs would cost 150 million rupees. Dr Sampooran Singh, director of the DRDO (Defence Research and Development Organisation) Terminal Ballistics Research Laboratory (TBRL) in Chandigarh, went so far as to write: 'A nuclear bomb programme is technically feasible, politically highly desirable, strategically inescapable and economically, not only sustainable, but actually advantageous. The cost of a nuclear weapons programme would, instead of crushing the Indian economy, accelerate growth of industry and technology. This would provide 50,000 jobs for engineers, scientists and technicians.'[23]

However, throughout the 1960s, no action was taken by the Indian government. The nuclear programme suffered severe setbacks when Prime Minister Shastri died in late 1965, followed by Dr Bhabha in an airline accident the following year. Shastri's successor, Indira Gandhi, was plagued with many internal and economic problems, and nuclear weapons consequently had a low priority. Moreover, there was a decrease in tension between India and China, meaning that the initial panic subsided. In addition, the military showed no major desire to possess nuclear weapons at that time. Yet while there was no overt weaponisation, the Indian government was extremely reluctant to relinquish its nuclear weapons option. The government of Indira Gandhi steadfastly refused to accede to the Treaty on the Non-Proliferation of Nuclear Weapons (NPT) which came into being in 1970, promoting the peaceful use of nuclear energy and global nuclear disarmament.

By the beginning of 1972, the basic design for India's first nuclear device was complete, and other parts of the programme for developing the necessary expertise to implement the design were coming along. During that year, the data from operating the Purnima nuclear research reactor (starting in May) began flowing in, allowing confirmation and refinement of the device's nuclear design. Meanwhile, work in plutonium metallurgy reached the point where the device could be successfully fabricated.

May 1972 saw the work of the Purnima reactor become critical. The reactor provided the basic data regarding neutron multiplication factors, effectiveness of reflectors and critical mass assembly, all of which were essential to the development of a workable nuclear weapon. Key issues such as how to achieve the optimum explosive power and the dependence of this optimum explosive power on the first self-sustaining nuclear trigger were investigated through a series of simulations undertaken by BARC scientists.[24] The Purnima facility used reprocessed plutonium from the 40MW CIRUS reactor and was of enormous help in the design and construction of India's first – and so far only – nuclear weapons test, which took place on 18 May 1974 at the Pokhran test site.

The 1974 Bomb

The leader of the team developing the Indian nuclear device was Raja Ramanna, director of the BARC. He worked with Dr Basanti Dulal Nag Chaudhuri, director of the DRDO and science advisor to the minister of defence, to coordinate the work of explosive system development. P.K. Iyengar and R. Chidambaram – who would come to prominence in 1998 – led the nuclear system design effort. In addition, Dr Nagapattinam Sambasiva Venkatesan directed the DRDO's TBRL, which developed and manufactured the high-explosive implosion system.[25]

The device tested at Pokhran had a yield of some 12 kilotons and used the implosion principle. It is not known exactly how much plutonium was used for the test, but some informed Indian estimates indicated that it was roughly 10kg – about one year's output from the Cirus reactor. This figure includes some plutonium lost during the machining of the core of the bomb, which leaves about 6–8kg in the device itself. The yield of the device was questioned, with estimates ranging from the official 12 kilotons to a low of just two kilotons, with the latter estimate being dismissed. However, P.K. Iyengar believed that the yield was 8–10 kilotons, based on radiochemical analysis from the shaft.[26]

The Pokhran device was exploded at a depth of 107 metres after being placed in an L-shaped hole. It was long believed that the device tested was a 'crude' bomb. However, in a 1994 magazine interview, the head of India's Atomic Energy Commission, Dr Rajagopala Chidambaram, stated that India was confident of the design prior to the test and almost boasted about how 'good' the Indian 'bomb' was.[27]

It is interesting to note that Dr Chidambaram, who was closely associated with designing the Pokhran device, used the word 'bomb' instead of the more widely used 'device'. Moreover, some observers have pointed out that the 1974 device was small enough to fit down a metal pipe and hence would be small enough to fit into a bomb casing. It may therefore be argued that the 1974 test was more of a weapon than a mere device.

The Indian civil nuclear programme suffered from a series of sanctions and safeguards imposed after the 1974 test. India, having refused to sign the NPT and having no agreements with the United States or Canada barring the use of materials in peaceful nuclear explosions – which is what the Pokhran test was called – did not violate any laws. It is a moot point just how 'peaceful' a nuclear explosion could be, but the United States did have a PNE programme dating to the late 1950s under the name 'Ploughshare'. Nevertheless, the Indian test did violate the spirit, if not the letter, of the Indo-Canadian agreement.

The impact on the Indian civil nuclear programme was substantial, and included the suspension, followed by the cancellation, of Canadian assistance for the construction of two nuclear power plants in Rajasthan, as well as the imposition of safeguards in perpetuity and 'pursuit' clauses upon the Rajasthan reactors for the supply of heavy water from the USSR.

The United States and France dramatically reduced supplies of enriched uranium for the Tarapur reactors and an 'Experimental Fast Breeder Test' reactor at Kalpakkam near Madras. The Nuclear Suppliers Group (popularly known as the London Club) was established in 1974 as a direct result of the Indian test in a bid to prevent nuclear proliferation, as was the 1978 passing of the US Nuclear Non-Proliferation Act. These measures are indicative of

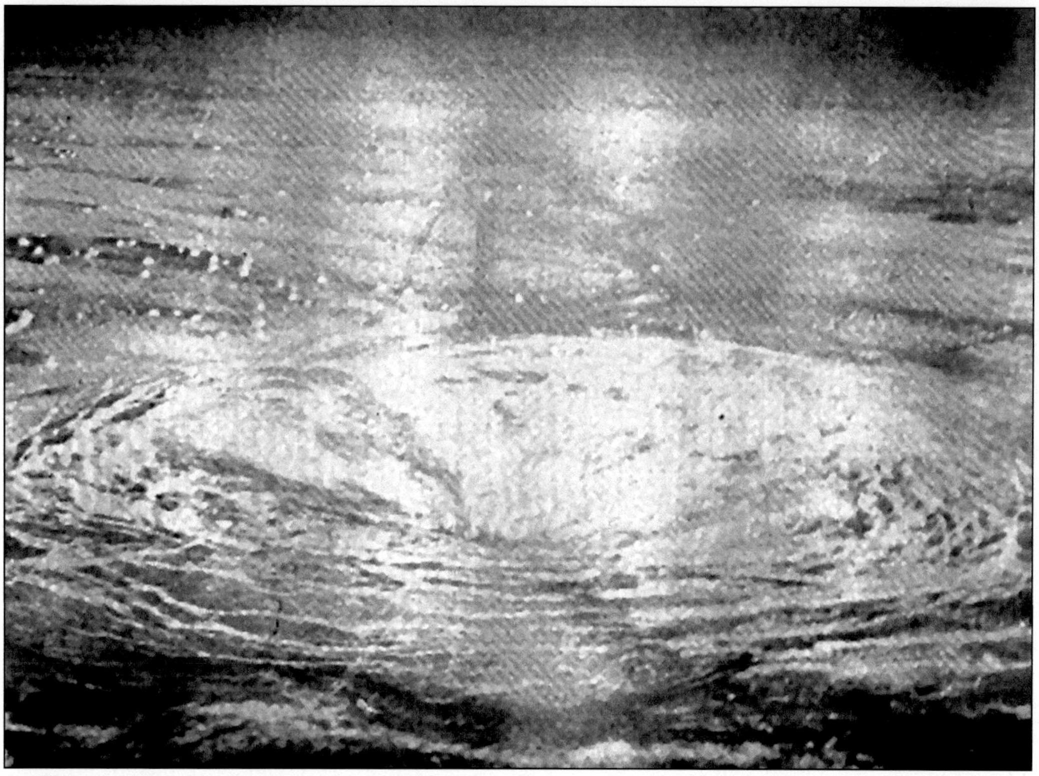

A still from a video showing the crater created by the underground test of the first Indian nuclear bomb, detonated on 18 May 1974. (PIB)

the alarm generated by India's nuclear test.[28]

Between 1977 and 1980, the Janata coalition government held power in India. Little is known about the state of the nuclear weapons programme during this time, and some have argued that the development project was allowed to stagnate. However, there is no clear evidence to suggest that this is the case, though it has been reported that the team that worked on the 1974 explosion was dispersed by the Janata government.

Upon Mrs Gandhi's return to power in 1980, it has been suggested that work on nuclear weapons was resumed. Evidence emerged that Pakistan had made rapid progress in the development of nuclear weapons in the mid-1980s, which must have accelerated India's nuclear weapons research. Rumours also circulated that India was prepared to conduct another test in 1984. Such preparations dated back to 1982/83, when Prime Minister Indira Gandhi, although fearful of US economic sanctions, approved nuclear tests in response to progress made by Pakistan in the nuclear sphere. The timidity of the Indian government was such that in order to secure the prime minister's approval, the scientists involved chose to present the tests to her as experiments, not the start of a test series of a weapons development programme. This proved insufficient to allow the tests to be carried out and also failed to prevent the prime minister from retracting her approval within hours of granting it, thus cancelling the nuclear tests scheduled for 1984.[29]

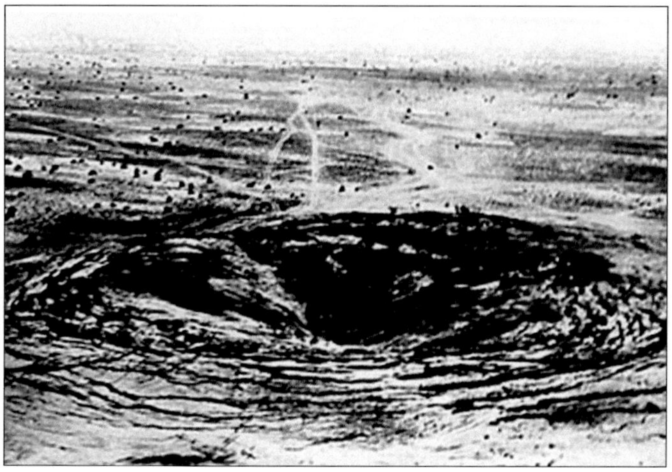

A close-up of the crater from the 1974 Indian nuclear weapons test. (PIB)

Nonetheless, India's nuclear infrastructure underwent a dramatic expansion during the 1980s. Several new power reactors were commissioned and a new research reactor was commissioned at the BARC. These, combined with existing plutonium reprocessing capabilities and the lack of international safeguards, substantially increased India's nuclear weapons design and production capabilities. These served alongside the CIRUS and Dhruva reactors, which were capable of producing substantial quantities of plutonium per year.

By 1990, four 235MWe pressurised heavy-water reactors (PHWR) were commissioned – two each at Madras in southern India and Narora in Uttar Pradesh. In addition, a 100MW research reactor – Dhruva – was commissioned in 1985 and, after numerous problems, reached full power by 1988.

Another view of the test-site of the Pokhran device, detonated at a depth of 107 metres beneath the ground. (PIB)

There is considerable evidence to indicate that India indulged in large-scale heavy-water smuggling to start these reactors. While such reports have been strenuously denied by the Indian authorities, others have also admitted that there was a significant shortfall in domestic heavy-water production in the early 1980s, the period during which the Dhruva research reactor and the two Madras power reactors were initially commissioned.[30] It is widely assumed that this smuggling was done to avoid international safeguards on the plutonium produced from these reactors. It should be noted that India apparently overcame its heavy-water production shortages and managed to obtain a sizeable surplus. In 1994, India obtained a contract to export 100 tonnes of heavy water to South Korea.

Work was also started in respect of thermonuclear and boosted fission weapons. India is believed to have programmes for both enhanced-fission and fusion weapons. The BARC conducts lithium 6 purification and production and has an extensive inertial confinement fusion (ICF) programme. This facility would be useful in the study of high-density physics associated with thermonuclear explosions, the improvement of elaborate computer codes and the development of sophisticated diagnostic techniques and instrumentation. Indeed, India may have completed all basic work on boosted-fission and fusion weapons by the 1980s.[31]

The War that Never Was and Kashmir 1990

In late 1986, India conducted a table-top exercise to explore new integrated mechanised warfare tactics. These evolved into full-scale troop exercises named Brasstacks, involving 150,000 men and thousands of tanks, armoured vehicles, artillery pieces and aircraft. It was this event, largely occasioned by misunderstanding and poor communications, that led to the crossing of the nuclear 'Rubicon' by both India and Pakistan.[32]

Pakistani support for Kashmiri terrorists also led to a confrontation in 1990, but this was on a smaller scale with fewer troop movements, though it was feared that escalation was possible. India's response to an armed Kashmiri insurgency – which resulted in the ethnic cleansing of almost the entire Hindu population from the valley of Kashmir – was harsh, brutal and uncompromising. It was during this confrontation that Pakistan began to hint at its nuclear weapons, using this to shield itself from Indian retaliation in the event of Pakistani-sponsored terror strikes. It was between 1987 and 1990 that both countries began the process of weaponising their hitherto nascent nuclear capability. Indeed, 1990 perhaps marked the point where Pakistan deployed its first nuclear weapons. This entire process culminated in the nuclear tests of 1998, when both nations demonstrated their nuclear weapons capability in the most dramatic fashion.

During the Brasstacks crisis, Prime Minister Rajiv Gandhi finally came to the realisation that Pakistan was either in possession of, or on the verge of, possessing the capability to build nuclear weapons. To this end, he – after much urging – authorised the DRDO to start development of rugged, miniaturised, safer and more reliable components and subsystems for what might eventually be a nuclear weapon system.

The Path to Weaponisation

However, even before the Brasstacks events, India's first effort to formulate a nuclear policy and to ascertain how to effectively implement it began in November 1985, aimed at answering questions posed by Rajiv Gandhi. Comprising representatives from the three services (Navy Chief of Staff Admiral Tahliani, Army Vice Chief of Staff General K. Sundarji and Deputy Chief of Air Staff John Greene) plus influential representatives from the BARC, DRDO and AEC, joined by India's most prominent strategic analyst, Krishnaswamy Subramanyam, this group suggested the creation of a minimum deterrent force of between 70 and 100 warheads, with a strict 'no first use' policy.[33] It appears that even at this time, a hastily cobbled-together system for weapons delivery was available.

While no direct action was taken in respect of the recommendations from that group, in 1986, Rajiv Gandhi instructed V.S. Arunachalam of the DRDO to deliver a system, with suitable control and security measures and improved reliability, to replace the stopgap system developed two years earlier. Arunachalam recruited K. Santhanam as technology adviser on the project, and the bomb system was codenamed 'New Armament Breaking Ammunition and Projectile', or NABAP.[34] Arunachalam went further and studied the basis for a basic command and control network that would be survivable. Based on his work, Arun Singh, an expert on strategic matters and senior former politician who had been Minister of State for Defence, was ordered to set up a national command post at a secure location near to New Delhi and had developed a semi-operational air-delivered deterrent.[35]

Unlike previous efforts, the Indian Air Force (IAF) was involved, which perhaps revealed flaws in the weapon's development, with that design being decidedly unsuitable for delivery from the Jaguar aircraft selected for delivery, only 2 inches of ground clearance being available.[36] By late 1986, the IAF rejected the Jaguar as unsuitable, and efforts switched to integrating the bomb with the recently acquired Mirage 2000. This led to the eventual adoption of this aircraft as the bomb's primary delivery system until missiles became available.

During the tensions over Kashmir in 1989 and 1990, India became acutely aware of its deficiencies in respect of nuclear command and control. To this end, the Arun Singh Committee was established to plan and prepare India's nuclear emergency response measures in the event of nuclear escalation. This committee prepared a series of emergency response procedures and established command and control mechanisms but did not deal with operational planning.

If Arun Singh Committee member Krishnaswamy Subramanyam is to be believed, the committee's 'only specific recommendation' was to 'to create separate storage for missiles and warheads ... what should be the drill for them being brought together ... and then ... the communications from command and control'. The American political scientist George Perkovich has claimed: 'The group called for designating air force units to receive nuclear devices and deliver them according to previously prepared orders that base commanders would possess under seal.'[37]

The 1990 crisis provided an additional spark for the redesigning of India's air-delivered bomb to ensure comfortable carriage under a Mirage 2000. This ultimately led to the certification of the air-delivery platform and weapon from a Mirage 2000 in 1994/95.[38] After this step, in 1995, Prime Minister Narasimha Rao approved the implementation of dispersal and concealment routines and

Table 1: Stockpiles of Indian Nuclear Materials 1991 and 1995

Production in kgs	Stockpile at end of 1991 (kgs)	Stockpile at end of 1995 (kgs)
Source		
Cirus reactor	245	280
Dhruva reactor	100	190
CANDU (First discharges)	15	30
Total Production	360	500
Consumption in kgs	**1991 in kgs**	**1995 in kgs**
1974 test	10	10
Processing losses (3%)	10	15
Fast reactor	50	50
Total consumption	70	75
Total Stockpile	290	425

procedures that had planned for safeguarding both the fissile cores and non-fissile trigger assemblies from a pre-emptive attack.[39]

Summary: Initial Weapons Capability Prior to the 1998 Tests

In the period leading up to the 1998 tests, the Indian plutonium stockpile had been estimated to be sufficient for 85 weapons, though this might have been an exaggeration.[40] India had developed a stockpile of some 15–20 nuclear gravity bombs in the period from 1974–85 – all of which were in the 15–20 kiloton range – and became an effective nuclear weapons state in May 1994 when it tested its air delivery system using Mirage 2000 aircraft. Though this could have been achieved much earlier, and indeed it was certainly possible for this to be done, hesitancy was the hallmark of early Indian nuclear weaponisation.

India had kept its weapons disassembled and un-mated to its delivery systems, anticipating that 72 hours would be sufficient to fully operationalise its deterrent. However, in the limited but very bitter Kargil conflict of 1999, precipitated when Pakistani troops and insurgents occupied Indian territory in the Kargil region – violating the Line of Control (LoC) to do so – India found itself having to prepare its weapons to deter any possible Pakistani nuclear first strike or demonstration of intent if the conflict escalated beyond the military actions taken to evict the intruding forces. Following shortly after the nuclear tests of 1998, the Kargil War brought up some old deficiencies in Indian nuclear planning and revived several issues of past years.

1
THE 1998 NUCLEAR TESTS

India, after prolonged development, has now deployed a large and sophisticated nuclear arsenal. In 1999, however, assessments of its nuclear capabilities were still largely confined to the view that only fission weapons in the form of free-fall bombs in the 15–20 kiloton range were available for deployment. With the benefit of hindsight, it is now quite well established that both India and Pakistan had deployed nuclear weapons in either the late 1980s or at the latest, the mid-1990s. What changed, however, was that in May 1998, India, then Pakistan, tested nuclear weapons, thus refining and experimenting with their designs as well as confirming their nuclear status.

India announced its arrival as a formal nuclear power with a rather brief statement from then Prime Minister Atal Bihari Vajpayee:

> Today at 1545 hrs, India conducted three underground nuclear tests in the Pokhran range. The tests were conducted with a fission device, a low yield device and a thermonuclear device. The measured yields are in line with expected values. Measurements have also confirmed that there was no release of radioactivity into the atmosphere. These were contained explosions like the experiment conducted in May 1974. I warmly congratulate the scientists and engineers who have carried out these successful tests.[1]

The rationale for these nuclear tests is sometimes questioned, but in a rather astonishing step, Prime Minister Vajpayee sent a series of letters to four of the five permanent members of the UN Security Council. The letter sent to then President Clinton was leaked in an apparent fit of pique by an angry US president. The text, which amply outlines India's rationale for testing in 1998, was as follows:

> Dear Mr. President,
> You would already be aware of the underground nuclear tests carried out in India. In this letter, I would like to explain the rationale for the tests.
>
> I have been deeply concerned at the deteriorating security environment, [e]specially the nuclear environment, faced by India for some years past. We have an overt nuclear weapon state on our borders, a state which committed armed aggression against India in 1962. Although our relations with that country have improved in the last decade or so, an atmosphere of distrust persists mainly due to the unresolved border problem. To add to the distrust that country has materially helped another neighbour of ours to become a covert nuclear weapons state. At the hands of this bitter neighbour we have suffered three aggressions in the last 50 years. And for the last ten years we have been the victim of unremitting terrorism and militancy sponsored by it in several parts of our country, [e]specially Punjab and Jammu & Kashmir. Fortunately, the faith of the people in our democratic system as also their patriotism has enabled India to counter the activities of the terrorists and militants aided and abetted from abroad.
>
> The series of tests are limited in number and pose no danger to any country which has no inimical intentions towards India. We value our friendship and cooperation with your country and you personally. We hope that you will show understanding of our concern for India's security.
>
> I assure you that India will continue to work with your country in a multilateral or bilateral framework to promote the cause of nuclear disarmament. Our commitment to participate in non-discriminatory and verifiable global disarmament measures is amply demonstrated by our adherence to the two conventions on Biological and Chemical Weapons. In particular we are ready to participate in the negotiations to be held in Geneva in the Conference on Disarmament for the conclusion of a fissile material cut-off treaty.
>
> I enclose for your information the text of the press statement issued after the nuclear tests were carried out today. I close with the expression of my highest consideration for your country and yourself.
>
> Yours sincerely,
> A.B. VAJPAYEE[2]

Surprising Surprise!

The US government declared itself to be surprised at India's multiple tests in May 1998. This is bizarre, since India had clearly been moving towards an openly declared nuclear capability for a number of years.

Table 2: The 1998 Weapon-Development Team Project Leaders	
Name	**Function**
Dr Avil (Abdul) Pakir Jainulabdeen Kalam	Scientific Adviser to the Defence Minister & Head of the Defence Research and Development Organisation (DRDO)
Dr Rajagopala Chidambaram	Chairman of India's Atomic Energy Commission (AEC) & Chairman of the Department of Atomic Energy (DAE)
Development and Test Teams for Nuclear Weapons	
Bhabha Atomic Research Centre (BARC) Leads	
Anil Kakodkar	Director of BARC
Satinder Kumar Sikka	Lead for Thermonuclear Weapon Development
M.S. Ramakumar	Director of Nuclear Fuel and Automation Manufacturing Group & Lead for Nuclear Component Manufacture
D.D. Sood	Director of Radiochemistry and Isotope Group & Lead for Nuclear Material Acquisition
S.K. Gupta	Solid State Physics and Spectroscopy Group; Device Design and Assessment
G. Goindraj	Associate Director of Electronic and Instrumentation Group; Lead for Field Instrumentation
DRDO Leads	
K. Santhanam	Lead for Test Site Preparations; Chairman of the (DAE)

Table 3: Test Preparations	
The Shakti Test Devices	
Designation	**Description**
Shakti I	Two-stage thermonuclear device with fusion-boosted primary, intended for missile warhead; test design yield 45kt, with a 200kt deployed yield
Shakti II	Lightweight pure fission tactical bomb/missile warhead, 12kt design yield
Shakti III	Fission experimental device, reportedly made with reactor-grade plutonium. Probably a fusion-boosted design without the fusion fuel, 0.3kt design yield
Shakti IV	0.5kt experimental device
Shakti V	0.2kt experimental device
Shakti VI	Not fired; another low-yield experimental device?

The rise of the Bharatiya Janata Party (BJP) – with its mix of populism, a degree of religious revivalism and chauvinism for the Hindu faith and its avowed nationalist agenda – placed to the forefront its manifesto of declaring India to be a nuclear power. A short-lived minority BJP government, which was formed for 13 days in May 1996, authorised nuclear tests; the devices got as far as being placed in the test shafts, but they were aborted when it was clear that the government would not last long enough to deal with the aftermath.

It was different, two years later, when on 10 March 1998, the BJP achieved a strong electoral victory and succeeded in putting together a governing coalition of 13 – later 20 – parties. A strong declaration of intent came on 18 March, the day before he was sworn in as prime minister, when Vajpayee declared: 'There is no compromise on national security. We will exercise all options including nuclear options to protect security and sovereignty.'[3] An official planning report further stated directly that the new BJP government intended to 're-evaluate the nuclear policy and exercise the option to induct nuclear weapons'.

This led to the preparation of a weapons development and testing team which had to undertake rapid preparations for nuclear tests, anticipating that they would not be able to repeat such tests as the Indian government anticipated a strong reaction from the United States. Moreover, the tests had to be done to preclude any pre-emptive and punitive economic actions against India on the part of the United States, and this involved a deception campaign to hide activity at the Pokhran test range as much as possible.

Ten days before the tests, the six Shatki devices were moved from their vaults at the BARC complex in Mumbai in the early morning hours of 1 May, around 3:00 a.m., by four Indian Army trucks under the command of Colonel Umang Kapur of the DRDO.[4] While the BARC and AEC dealt with the construction of the fissile elements of the weapons, three laboratories of the DRDO were engaged to design, test and produce non-fissile but nonetheless essential components such as advanced detonators, the implosion systems and high-voltage trigger systems. They were also responsible for weaponisation – systems engineering, aerodynamics, safety interlocks and flight trials, which earlier cleared air-delivered weapons and later produced useable missile warheads for the Agni and Prithvi missiles which were emerging as primary delivery systems for India's warheads.[5]

Each fissile core, or the plutonium for the devices, weighed 3–8kg, depending on the device. Witnesses noted that the cores were a grey colour due to the coating applied to contain the radioactivity and also to prevent oxidation of the plutonium. In contrast, the explosives surrounding the cores were a dull orange colour.

The tests were conducted in two batches – the first consisted of the thermonuclear device, the fission bomb and a sub-kiloton device, while two more sub-kiloton devices made up the second batch. In each case, all devices were detonated simultaneously, which created some level of yield uncertainty which persists to this day in academic circles.

Yields and Disputes

Known as Operation Shakti-98, the five tests had a recorded seismic magnitude of 5.0 +/- 0.4 on the Richter scale.[6] By 14 May, Indian seismologists at the Bhabha Atomic Research Centre stated that an analysis of the data received at the Gauribidnaur Seismic Array indicated that the yield of the 11 May blasts was around 55 kilotons.[7] They also indicated that the seismic waveform was very complex because the explosions were carried out simultaneously.

The first group of tests – Shakti I, II and III – consisted of a two-stage thermonuclear device with a yield of 43–45kt, a lightweight fission device with a yield of 12–15kt and a low-yield sub-kiloton device of 0.2kt. These were placed in two shafts, 1km apart, and simultaneously detonated from a control located 3.5km away.[8] Shakti IV and V had yields of 0.5kt and 0.3kt and were conducted to provide additional data for improved computer simulation of designs.

Almost immediately, a somewhat unseemly row erupted between Indian scientists and certain American seismologists regarding the yields of the devices. Preliminary estimates put the combined yields at anywhere between 10kt and 25kt, and some questioned whether India had actually tested a thermonuclear device based on these yield estimates. This row continued, with an American scientist, Dr Terry Wallace, giving a yield of 10–15kt and practically accusing the Indians scientists of lying.[9]

According to David Albright, a scientist at the Institute of Science and International Security in Washington DC, the lower estimates were largely due to the fact that a few Western scientists were of the view that India's first atomic test in 1974 fizzled out and produced a yield of only 2kt.[10] However, P.K. lyengar, former chairman of India's Atomic Energy Commission, has stated that the yield was between 8kt and 10kt, based on radiochemical analysis of samples of bomb debris taken from the shaft.[11]

Soon after Wallace's criticism, three Indian scientists writing in *Current Science* gave a detailed account of why Wallace was wrong. These scientists pointed out that there was a very large variation in the estimates of the body-wave magnitude made by different seismic stations. The variations at stations at similar distances from the explosion site varied by a staggering factor of 30 – or three times as much as expected.[12] The scientists pointed out that seismic stations located to the east and west of Pokhran recorded lower magnitudes than those to the north and south. Since two large explosions – the thermonuclear device and the fission device – were detonated in separate shafts 1km apart in an east–west direction, they said delays between the two signals could result in seismic stations recording lower strengths than the actual values.[13]

The *Current Science* article went on to say that seismic stations to the north and south of Pokhran would have the least time delay, and so would record the highest signal strength, while those to the east and west would have the maximum delay. Averaging out all seismic stations would therefore lower the value. Moreover, the true body-wave magnitude would have to be obtained by superimposing the signals from the two powerful explosions.[14]

The three scientists – Dr S.K. Sikka, Dr Falguni Roy and Dr G.K. Nair of the DARC High Pressure Physics Division – believed that a true estimate of the body-wave magnitude could only be obtained by taking the data from seismic stations to the north and south of Pokhran. The data from 10 seismographs to the north and two to the south indicated a body-wave magnitude of 5.36. If the reading of one station was excluded because the signal-to-noise ratio was too low, the average value rose to 5.39, which tallied with the BARC's own measurements at Gauribidanaur. This data was in agreement with the yield claims made by Indian weapons scientists.[15]

To complicate matters further, Dr Roger Clark, a seismologist at Leeds University, stated that after data from 12 seismic stations was taken into account, the yield estimate was close to 60kt. Furthermore, the US Geological Survey indicated that the body-wave magnitude was 5.2, which suggested a yield of 30–60kt.[16]

Of great concern was the fact that the 13 May explosions were undetected. This has led to scientists such as Wallace alleging that these tests never actually took place. However, Dr Clark points out that if the Indians tested in a sand dune, as claimed by Dr Chidambaram, or if the devices were exploded in a large cavity, then the tests could have been completely concealed. The other possibility was that the tests were hydronuclear or sub-critical, as was claimed by Dr Raja Ramanna of the BARC.[17]

Another major controversy over India's nuclear tests was that there was considerable doubt that India had detonated a thermonuclear device. Once the yield estimates were more or less verified, doubts focused on the fact the device seemed to have a low yield. The yield of 43kt was said by some to be too small for a thermonuclear device, and they questioned whether India's device had worked as planned.

This is factually incorrect. The United States, for example, tested two thermonuclear devices in 1953, with yields of 23kt and 43kt. Dr Frank Barnaby, writing in *Jane's Defence Weekly*, also disagreed, stating that thermonuclear devices did not need to be tested at full yield and could be limited to firing the nuclear-fission trigger and producing a small amount of fusion energy.[18] It will be shown that India had all the components in place for a thermonuclear design. As such, scepticism in this regard seems unwarranted.

Dr Barnaby suggested that an operational Indian thermonuclear weapon was likely to contain a boosted-fission nuclear trigger giving an explosion of 50kt, which would produce 500kt in the fusion stage.[19] He further suggested that a variable-yield nuclear weapon could be produced in this way, allowing a yield of 5kt, 50kt or 500kt to be chosen.[20] One Indian nuclear scientist claimed that the device tested was a 'third-generation' fusion device.

It should also be noted that even if lower estimates are accepted – with verifiable Indian drilling data and crater analysis putting a lowest yield of the Shakti I device at 22kt – this is at or above the expected yield range of a thermonuclear primary, making it evident that the fusion-boosted device used for the primary was successful. Furthermore, the Shakti III yield of 0.3kt is appropriate for the unboosted yield of a boosted fission design – in other words, a device that is fired without the deuterium-tritium 'boosting' that would increase its yield. This indicates that India demonstrated the ability to develop and employ full-yield fusion-boosted weapons using less-than-weapons-grade plutonium, a fact which has major significance for India's current arsenal.[21]

Table 4: Indian Nuclear Capability in 1999[22]

Reactor	Capacity	Plutonium Output	Weapons Potential
CIRUS	40MW	9–10kg/year	1–2/year
Dhruva	100MW	20–25kg/year	4–5/year
MAPP – Madras	2 x 235MWe	2 x 60kg/year	2 x 12 = 24/year
NAPP – Narora	2 x 235MWe	2 x 60kg/year	2 x 12 = 24/year
KAPP – Kakrapar	2 x 235MWe	2 x 60kg/year	2 x 12 = 24/year
Total		389–395kg/year	77–79kg/year

This latter point cannot be overstated, as even in 1998/9, India's nuclear capability rested on the potential use of the reactors listed in Table 4.

It should be noted that the power plants at Madras, Narora and Kakrapar may or may not have been used for the manufacture of weapons-grade plutonium. However, they were definitely capable of such production, and unsafeguarded reactors of this type remain a potential factor in calculations of India's nuclear weapons potential. Whether India used any of these power reactors for weapons production is debatable, but as it is strongly suspected that reactor-grade plutonium was used in at least some of the Shatki series tests to determine whether it could be usefully used to bolster fissile material stocks for weapons production, and given even the most sceptical interpretation of the results of the test, this capability and the capability to produce full-yield boosted-fission weapons was undoubtedly demonstrated.

Weaponisation – Developing and Deploying the Deterrent in the 1990s

Neither India nor Pakistan has publicly discussed its nuclear weapons deployment during the Kargil conflict. However, it is known, that India readied four Prithvi SRBMs (Short-Range Ballistic Missiles) and one Agni IRBM (Intermediate-Range Ballistic Missile) for possible use in a retaliatory role in the event of the use of a Pakistani weapon, as well as preparing some of its air force Mirage 2000s to delivery free-fall bombs, bringing its arsenal to Readiness State 3, with warheads ready to be mated at short notice.[23]

India's first deployed air-delivered nuclear weapons were earmarked for use from the Jaguar strike aircraft. One would have assumed that in the 1980s and into the early 1990s, the Jaguar – with an excellent navigation-attack system and a very good payload capacity – would have been an ideal choice. However, the type's less-than-stellar flight performance combined with practical problems of ground clearance, led to the Mirage 2000 being chosen to perform this task, with active trials of the system being undertaken in 1994.

Initial tests with the Jaguar IS – India's principal deep-penetration strike aircraft – found that the ARDE (Armament Research and Development Establishment) designed bomb-pods were too heavy for the Jaguar, and once fitted to the centreline pylon of the aircraft, the ground clearance of a mere 2 inches was deemed unsafe.[24] India thus chose to use its Mirage 2000H/TH fleet for the nuclear delivery role.

It was with this aircraft that India finally completed its development of a fully combat-ready and deployable system for delivering nuclear weapons, with the IAF conducting acceptance trials. The ARDE-developed bomb case and the TBRL-developed implosion system were mated with a modified Mirage 2000 and successfully test-dropped at Balasore using a toss-bombing technique. The bomb, minus its plutonium core, was fused for an airburst and released over the Bay of Bengal. With this test, India had a reasonably reliable delivery system. It also now had an arsenal of at least a couple of dozen operational nuclear bombs.[25]

Development of India's air-launched deterrent was somewhat chaotic. However, at the urging of Army Chief of Staff General Sundarji, Prime Minister Rajiv Gandhi had instructed the DRDO to start development of a rugged, miniaturised and safer nuclear weapon and to develop more reliable components and subsystems to 'keep the country's nuclear capability at least at a minimum state of readiness'.[26]

While some argue that this stopped short of ordering the building of a weapon or integrating it into a delivery platform, it could be

Prime Minister Rajiv Gandhi began the process towards weaponisation. (PIB)

Prime Minister Narasimha Rao completed the weaponisation process. (PIB)

Prime Minister Atal Bihari Vajpayee tested new weapons designs in 1998 and operationalised the nuclear arsenal. (PIB)

debated *ad nauseum* as that period of Indian nuclear history is still somewhat unclear. However, eventually, in 1989, weaponization was approved after Rajiv Gandhi's global disarmament plan met with abject failure.[27] In the wake of the failure of this plan, menacing Indian intelligence reports concluded in March 1988 that 'Pakistan was in possession of at least three nuclear devices of 15–20 kiloton yield'.[28]

No missiles were yet in the offing, and the instructions from India's prime minister were to concentrate on the development and certification of air-deliverable weapons and the creation of an aircraft delivery package that was both safe and reliable.[29] This was done, but it took several years and a new prime minister – in the form of P.V. Narasimha Rao – to complete this task.

Whether true or not, it is widely believed that up to 1998, the Indian Air Force was the only military service with any knowledge of India's nuclear weapons programme, and that was only because of its role in delivering the weapons.[30] While there was much unofficial knowledge of the fact that weapons were available for use, international security expert Professor Gaurav Kampani noted that:

> A senior Indian defense official privy to this effort disclosed that, until 1999, the air force had no idea (1) what types of weapons were available; (2) in how many numbers; and (3) what it was expected to do with the weapons. All the air force had was delivery capability in the form of a few modified Mirage 2000s.[31]

In fact, events at Kargil were to be an eye-opening experience, where the Indian nuclear deterrent took nearly a week to achieve any degree of capability – transitioning from recessed deterrent to employment mode – as opposed to an earlier calculation of just 72 hours.[32] Indeed, Kampani notes that the IAF was not ready until the end of June 1999 to commence nuclear operations against Pakistan if ordered.[33]

However, the IAF was not the only agency involved, as the DRDO had readied one Agni and four Prithvi missiles for the nuclear strike role against Pakistan, with a trajectory for the former being worked out so that its discarded stages did not fall in Indian territory. Activation of this Readiness State 3, while taking longer than anticipated, marked the deployment of an operational air- and missile-based deterrent for India.[34]

That these preparations were, in fact, warranted can be gleaned from a 2018 statement by Bruce Riedel, a former CIA analyst and White House aide during the Kargil War, who indicated that the CIA's daily brief for 4 July 1999 indicated that Pakistan was preparing its nuclear weapons for deployment and possible use.[35] In the event, neither India nor Pakistan overtly threatened the use of their nuclear weapons during this conflict, and it is possible that the plausible deniability sought by Pakistan, plus India's response in not crossing the LoC, were both important factors in moderating nuclear rhetoric.

2
THE INDIAN DOCTRINE – NO FIRST USE[1]

The Indian nuclear doctrine as detailed in a press release on 4 January 2003 is a starkly simplistic document. Indeed, the cabinet note from that meeting spoke of 'operationalising' the nuclear doctrine. The 2003 doctrine derives from a much more detailed draft doctrine of 1999 and these two documents must be read together and cannot be seen in isolation. India is one of the only nuclear powers that has opened its nuclear doctrine, in a draft form, for scrutiny and study prior to its adoption.

On 17 August 1999, the then Indian national security advisor Brajesh Mishra released a draft report from the National Security Advisory Board on Indian Nuclear Doctrine. This report, which outlines in broad terms India's rationale and intentions regarding the development of its 'minimum nuclear deterrent', was never formally approved by the 1999 government of then Prime Minister Vajpayee. However, given the brevity of the 2003 doctrine, the rationale therein outlined can serve as an important guide to the reasons for the approach India has taken to the doctrine. The two doctrines are here replicated in full.

Draft Indian Nuclear Doctrine

1. **Preamble**

   1.1. The use of nuclear weapons in particular as well as other weapons of mass destruction constitutes the gravest threat to humanity and to peace and stability in the international system. Unlike the other two categories of weapons of mass destruction, biological and chemical weapons which have been outlawed by international treaties, nuclear weapons remain instruments for national and collective security, the possession of which on a selective basis has been sought to be legitimised through permanent extension of the Nuclear Non-proliferation Treaty (NPT) in May 1995. Nuclear weapon states have asserted that they will continue to rely on nuclear weapons with some of them adopting policies to use them even in a non-nuclear context. These developments amount to virtual abandonment of nuclear disarmament. This is a serious setback to the struggle of the international community to abolish weapons of mass destruction.

   1.2. India's primary objective is to achieve economic, political, social, scientific and technological development within a peaceful and democratic framework. This requires an environment of durable peace and insurance against potential risks to peace and stability. It will be India's endeavour to proceed towards this overall objective in cooperation with the global democratic trends and to play a constructive role in advancing the international system toward a just, peaceful and equitable order.

   1.3. Autonomy of decision making in the developmental process and in strategic matters is an inalienable democratic right of the Indian people. India will strenuously guard this right in a world where nuclear weapons for a select few are sought to be legitimised for an indefinite future, and where there is growing complexity and frequency in the use of force for political purposes.

1.4. India's security is an integral component of its development process. India continuously aims at promoting an ever-expanding area of peace and stability around it so that developmental priorities can be pursued without disruption.

1.5. However, the very existence of offensive doctrine pertaining to the first use of nuclear weapons and the insistence of some nuclear weapons states on the legitimacy of their use even against non-nuclear weapon countries constitute a threat to peace, stability and sovereignty of states.

1.6. This document outlines the broad principles for the development, deployment and employment of India's nuclear forces. Details of policy and strategy concerning force structures, deployment and employment of nuclear forces will flow from this framework and will be laid down separately and kept under constant review.

2. Objectives

2.1. In the absence of global nuclear disarmament India's strategic interests require effective, credible nuclear deterrence and adequate retaliatory capability should deterrence fail. This is consistent with the UN Charter, which sanctions the right of self-defence.

2.2. The requirements of deterrence should be carefully weighed in the design of Indian nuclear forces and in the strategy to provide for a level of capability consistent with maximum credibility, survivability, effectiveness, safety and security.

2.3. India shall pursue a doctrine of credible minimum nuclear deterrence. In this policy of 'retaliation only', the survivability of our arsenal is critical. This is a dynamic concept related to the strategic environment, technological imperatives and the needs of national security. The actual size components, deployment and employment of nuclear forces will be decided in the light of these factors. India's peacetime posture aims at convincing any potential aggressor that:

(a) any threat of use of nuclear weapons against India shall invoke measures to counter the threat: and

(b) any nuclear attack on India and its forces shall result in punitive retaliation with nuclear weapons to inflict damage unacceptable to the aggressor.

2.4. The fundamental purpose of Indian nuclear weapons is to deter the use and threat of use of nuclear weapons by any State or entity against India and its forces. India will not be the first to initiate a nuclear strike, but will respond with punitive retaliation should deterrence fail.

2.5. India will not resort to the use or threat of use of nuclear weapons against States which do not possess nuclear weapons, or are not aligned with nuclear weapon powers.

2.6. Deterrence requires that India maintain:

(a) Sufficient, survivable and operationally prepared nuclear forces,

(b) a robust command and control system,

(c) effective intelligence and early warning capabilities,

(d) comprehensive planning and training for operations in line with the strategy, and

(e) the will to employ nuclear forces and weapons.

2.7. Highly effective conventional military capabilities shall be maintained to raise the threshold of outbreak both of conventional military conflict as well as that of threat or use of nuclear weapons.

3. Nuclear Forces

3.1. India's nuclear forces will be effective, enduring, diverse, flexible and responsive to the requirements in accordance with the concept of credible minimum deterrence. These forces will be based on a triad of aircraft, mobile land-based missiles and sea-based assets in keeping with the objectives outlined above. Survivability of the forces will be enhanced by a combination of multiple redundant systems, mobility, dispersion and deception.

3.2. The doctrine envisages assured capability to shift from peacetime deployment to fully employable forces in the shortest possible time, and the ability to retaliate effectively even in a case of significant degradation by hostile strikes.

4. Credibility and Survivability

The following principles are central to India's nuclear deterrent:

4.1. Credibility: Any adversary must know that India can and will retaliate with sufficient nuclear weapons to inflict destruction and punishment that the aggressor will find unacceptable if nuclear weapons are used against India and its forces.

4.2. Effectiveness: The efficacy of India's nuclear deterrent be maximised through synergy among all elements involving reliability, timeliness, accuracy and weight of the attack.

4.3 Survivability:

(i) India's nuclear forces and their command and control shall be organised for very high survivability against surprise attacks and for rapid punitive response. They shall be designed and deployed to ensure survival against a first strike and to endure repetitive attrition attempts with adequate retaliatory capabilities for a punishing strike which would be unacceptable to the aggressor.

(ii) Procedures for the continuity of nuclear command and control shall ensure a continuing capability to effectively employ nuclear weapons.

5. Command and Control

5.1. Nuclear weapons shall be tightly controlled and released for use at the highest political level. The authority to release nuclear weapons for use resides in the person of the Prime Minister of India, or the designated successor(s).

5.2. An effective and survivable command and control system with requisite flexibility and responsiveness shall be in place. An integrated operational plan, or a series of sequential plans, predicated on strategic objectives and a targeting policy shall form part of the system.

5.3. For effective employment the unity of command and control of nuclear forces including dual capable delivery systems shall be ensured.

5.4. The survivability of the nuclear arsenal and effective command, control, communications, computing, intelligence and information (C4I2) systems shall be assured.

5.5 The Indian defence forces shall be in a position to execute operations in an NBC environment with minimal degradation.

5.6. Space based and other assets shall be created to provide early warning, communications, damage/detonation assessment.

6. **Security and Safety**

6.1. Security: Extraordinary precautions shall be taken to ensure that nuclear weapons, their manufacture, transportation and storage are fully guarded against possible theft, loss, sabotage, damage or unauthorised access or use.

6.2. Safety is an absolute requirement and tamper proof procedures and systems shall be instituted to ensure that unauthorised or inadvertent activation/use of nuclear weapons does not take place and risks of accident are avoided.

6.3. Disaster control: India shall develop an appropriate disaster control system capable of handling the unique requirements of potential incidents involving nuclear weapons and materials.

7. **Research and Development**

7.1. India should step up efforts in research and development to keep up with technological advances in this field.

7.2. While India is committed to maintain the deployment of a deterrent which is both minimum and credible, it will not accept any restraints on building its R&D capability.

8. **Disarmament and Arms Control**

8.1. Global, verifiable and non-discriminatory nuclear disarmament is a national security objective. India shall continue its efforts to achieve the goal of a nuclear weapon-free world at an early date.

8.2. Since no first use of nuclear weapons is India's basic commitment, every effort shall be made to persuade other States possessing nuclear weapons to join an international treaty banning first use.

8.3. Having provided unqualified negative security assurances, India shall work for internationally binding unconditional negative security assurances by nuclear weapon states to non-nuclear weapon states.

8.4. Nuclear arms control measures shall be sought as part of national security policy to reduce potential threats and to protect our own capability and its effectiveness.

8.5. In view of the very high destructive potential of nuclear weapons, appropriate nuclear risk reduction and confidence building measures shall be sought, negotiated and instituted.[2]

Following this draft, a Press Information Bureau release of 4 January 2003 gave the first indication of what elements of the draft nuclear doctrine had been agreed and which ones had either been omitted or changed somewhat:

CABINET COMMITTEE ON SECURITY REVIEWS PROGRESS IN OPERATIONALIZING INDIA'S NUCLEAR DOCTRINE

1. The Cabinet Committee on Security (CCS) met today to review the progress in operationalizing of India's nuclear doctrine. The Committee decided that the following information, regarding the nuclear doctrine and operational arrangements governing India's nuclear assets, should be shared with the public.
2. India's nuclear doctrine can be summarized as follows:
   i. Building and maintaining a credible minimum deterrent;
   ii. A posture of 'No First Use' nuclear weapons will only be used in retaliation against a nuclear attack on Indian territory or on Indian forces anywhere;
   iii. Nuclear retaliation to a first strike will be massive and designed to inflict unacceptable damage;
   iv. Nuclear retaliatory attacks can only be authorised by the civilian political leadership through the Nuclear Command Authority;
   v. Non-use of nuclear weapons against non-nuclear weapon states;
   vi. However, in the event of a major attack against India, or Indian forces anywhere, by biological or chemical weapons, India will retain the option of retaliating with nuclear weapons;
   vii. A continuance of strict controls on export of nuclear and missile related materials and technologies, participation in the Fissile Material Cutoff Treaty negotiations, and continued observance of the moratorium on nuclear tests;
   viii. Continued commitment to the goal of a nuclear weapon free world, through global, verifiable and non-discriminatory nuclear disarmament.
3. The Nuclear Command Authority comprises a Political Council and an Executive Council. The Political Council is chaired by the Prime Minister. It is the sole body which can authorize the use of nuclear weapons.
4. The Executive Council is chaired by the National Security Advisor. It provides inputs for decision making by the Nuclear Command Authority and executes the directives given to it by the Political Council.
5. The CCS reviewed the existing command and control structures, the state of readiness, the targeting strategy for a retaliatory attack, and operating procedures for various stages of alert and launch. The Committee expressed satisfaction with the overall preparedness. The CCS approved the appointment of a Commander-in-Chief, Strategic Forces Command, to manage and administer all Strategic Forces.
6. The CCS also reviewed and approved the arrangements for alternate chains of command for retaliatory nuclear strikes in all eventualities.[3]

Differences between the Draft Doctrine and the Operationalised Doctrine

The National Security Advisory Board prepared its Draft Nuclear Doctrine suggesting 'massive retaliation' for consideration by the government in August 1999, shortly after the end of the Kargil War, with the relevant portion stating:

> Any nuclear attack on India and its forces shall result in punitive retaliation with nuclear weapons to inflict damage unacceptable to the aggressor [Para 2.3 (b)].
>
> In the aftermath of the military mobilization of 2001–02 as part of Operation Parakram, the Cabinet Committee on Security on January 4, 2003, issued a press release which adopted the nuclear doctrine as detailed above. Where it differs from the draft doctrine, in operative terms is as follows:
>
> '(ii) A posture of "No First Use": Nuclear weapons will only be used in retaliation against a nuclear attack on Indian territory or on Indian forces anywhere;
>
> (iii) Nuclear retaliation to a first strike will be massive and designed to inflict unacceptable damage.'[4]

Some analysts rightly assert that the doctrine was now one of 'assured retaliation' to nuclear weapons use by a hostile power, and in the event of a major 'first strike', this retaliation would be massive. At 'sub-first-strike levels', the retaliation may be graduated or of a lower order, which suggested a 'flexible' or 'graduated' response.

These subtle shifts in terminology were not merely semantic, having as they did significant meaning. In the chapter on tactical nuclear weapons, it was noted that the Indian nuclear doctrine seems to have inherent flexibility built in, with some scholars noting that a tactical strike on a military formation in enemy territory does not fall within the ambit of a 'first strike'. This suggests that with this change of wording, India had moved away from the concept of 'massive retaliation' towards one of a 'flexible response'.[5]

The question that is unknown is whether India itself believes that it has now adopted a 'flexible response' doctrine or is still thinking in terms of 'massive retaliation'. The question as to what India's retaliation would be to sub-first-strike attacks is open to question. That 'massive retaliation' seems to be limited to a response to a 'first strike' suggests that a response to a lower-level attack could be 'flexible'.

There has been some debate as to whether India is intending to abandon its NFU (No First Use) doctrine. There are several factors which may make this less than desirable, primarily focusing on international reaction and complicated command and control.[6]

India adopting a first-use doctrine against a larger nuclear power such as China would be impractical, not to mention unnecessarily provocative without any strategic benefit, while a first-use doctrine against Pakistan is unnecessary in any circumstance. Ambiguity over possible first use at this stage would be counterproductive, as a change from the NFU would be seen as adopting a potential first-use doctrine. To this end, it is suggested that the policy of NFU be retained and revisions, if any, to the Indian doctrine be confined to clarifying ambiguities.

Having said that, this imposes some requirements upon the Indian deterrent, and the progress towards it will determine whether the NFU doctrine is credible, viable and sustainable. These include issues of command and control, the numerical size and nature of the warheads, their readiness levels and, most importantly, the survivability of a large-enough force to ensure a level of unacceptable destruction on any adversary.[7]

A SEPECAT Jaguar International – designated *Shamsher* in Indian Air Force service – seen releasing a pair of 1,000lb (454kg) bombs. (PIB)

The new Jaguar upgrade fits an ELTA 2052 radar to the Jaguar and will give some 60 aircraft dramatically enhanced capabilities. (PIB)

The NFU Debate – Much Ado About Nothing[8]

In March 2017, Professor Vipin Narang of the Department of Political Science at the Massachusetts Institute of Technology made the somewhat startling claim that

India was rethinking its 'no-first-use' nuclear doctrine. In an address delivered to the Carnegie International Nuclear Policy Conference, Professor Narang made the following statement:

> There is increasing evidence that India will not allow Pakistan to go first. And that India's opening salvo may not be conventional strikes trying to pick off just Nasr batteries in the theater, but a full 'comprehensive counter-force strike' that attempts to completely disarm Pakistan of its nuclear weapons so that India does not have to engage in iterative tit-for-tat exchanges and expose its own cities to nuclear destruction. This thinking surfaces not from fringe extreme voices such as Bharat Karnad or retired Indian Army officers frustrated by the lack of resolve they believe their government has shown in multiple provocations, but from no less than a former Strategic Forces Command C-in-C Lt Gen BS Nagal and, perhaps more importantly and authoritatively, from the highly respected and influential former National Security Advisor Shivshankar Menon in plain sight in his recent 2016 book *Choices: Inside the Making of Indian Foreign Policy*. In short, we may be witnessing what I call a 'decoupling' of Indian nuclear strategy between China and Pakistan. The force requirements India needs in order to credibly threaten assured retaliation against China may allow it to pursue more aggressive strategies – such as escalation dominance or a 'splendid first strike' – against Pakistan. We may be seeing the emergence of this decoupling, or at least serious mainstream thinking about it, with the intention being a disarming strike against Pakistan.[9]

If this is true, it would indeed mark a substantial and complete shift in India's nuclear doctrine, which had hitherto been assumed to be firmly entrenched in the NFU approach to nuclear weapons. Professor Narang derived his postulations, at least in part, from a paragraph in a book written by former Indian National Security Advisor Shivshankar Menon which states:

> If Pakistan were to use tactical nuclear weapons against India, even against Indian forces in Pakistan, it would effectively be opening the door to a massive Indian first strike, having crossed India's declared red lines. There would be little incentive, once Pakistan had taken hostilities to the nuclear level, for India to limit its response, since that would only invite further escalation by Pakistan. India would hardly risk giving Pakistan the chance to carry out a massive nuclear strike after the Indian response to Pakistan using tactical nuclear weapons. In other words, Pakistani tactical nuclear weapons use would effectively free India to undertake a comprehensive first strike against Pakistan.[10]

However, there is a significant caveat in accepting Narang's conclusions from this paragraph. While the words 'comprehensive first strike' are used, it is clear from the context of the paragraph that Menon does not mean any sort of nuclear first strike. Indeed, it suggests the exact opposite, with Menon meaning 'second strike' or some equivalent pre-emptive non-nuclear strike. This latter interpretation would be in accord with the remainder of the chapter in which the offending paragraph is to be found, where Menon embarks upon a rigorous defence of India's NFU approach.[11]

This is especially so since doctrines cannot be seen in isolation and must be set against the path India has taken in developing its force structure. With its emphasis on mobility and survivability, it is suggested that there is no evidence of a shift away from an NFU doctrine on the part of India. Furthermore, it is suggested that India has always practically decoupled its approach to Pakistan and China owing to the latter's NFU doctrine and the former's propensity to relish in its rejection of the same. It is thus submitted that Narang may be exaggerating to an extent.

With clarity on neither NFU nor a first-strike approach, the possibility of a doctrine of ambiguity destabilising what strategic balance might exist would be a distinct possibility. Furthermore, a shift to a policy of ambiguity, after having acknowledged adhering to one of NFU, could negate whatever diplomatic credits might have accrued to India through its adoption of the NFU doctrine. In addition, any overt shift would create wholly understandable alarm in Beijing owing to China's own NFU doctrine.

A pair of Mirage 2000H Vajras from No. 1 Squadron, Indian Air Force. (IAF)

A still from a video showing a Mirage 2000H Vajra armed with the Israeli-made Spice guided bomb. (IAF)

In such circumstances, it is difficult to discern what advantage – either operationally, strategically or diplomatically – might be obtained from a shift in India's consistent assertion of its NFU doctrine. However, a much greater concern for all observers of India's nuclear doctrine, as has been alluded to before, is whether there is a consistent approach to India's nuclear weapons doctrine in both word and practice.

A more significant issue, as indicated earlier in this chapter, is whether those articulating NFU have a clear understanding of the nuances therein. In nuclear issues, phraseology is important. One particular paragraph in Menon's book could potentially imply that India has not ruled out a strike before an adversary actually *detonates* a nuclear weapon against an Indian target:

> There is a potential gray area as to when India would use nuclear weapons first against another NWS [nuclear weapons state]. Circumstances are conceivable in which India might find it useful to strike first, for instance, against an NWS that had declared it would certainly use its weapons, and if India were certain that adversary's launch was imminent.[12]

On the face of it, a literal interpretation of India's nuclear doctrine would suggest that the above scenario was precluded by NFU. However, that Menon chose to raise such a case might imply that there has been some consideration given to the matter. It must be stressed, however, that *certainty* of an adversary's launch would be an extremely difficult proposition for India to ensure. Even with elaborate provision for launch detection and effective intelligence of an adversary's intention, the practical difficulties in guaranteeing that, for example, an enemy missile launch is carrying a nuclear warhead should preclude abrogating the NFU doctrine in such circumstances.

It does not, however, preclude the use of conventional weapons to destroy an adversary's delivery systems before launch. While India still lacks sufficient assets to ensure a completely effective conventional counter-force strike, it is not inconceivable that India could, without perhaps undue difficulty, target munitions stockpiles and individual mobile missile launchers as part of a broader military campaign with, in the case of the short-range Nasr system, a fair chance of success.

The Credible Minimum Deterrent Concept[13]

One of the cornerstones of the Indian Nuclear Doctrine is the concept of the Credible Minimum Deterrent. This inevitably requires a number of components:

> Sufficient and survivable nuclear forces both in terms of warheads and means of delivery able to inflict unacceptable damage
>
> Nuclear forces must be operationally prepared at all times
>
> Effective intelligence and early-warning capabilities
>
> A robust command and control system
>
> The Will to Employ Nuclear Forces
>
> Communication of deterrence capability

The question as to how far India has come along the path to having a credible deterrent, as well as a viable doctrine that merges into a coherent strategy in the event of a conflict, is open to some debate. The reticence of Indian officials to comment on the state of the nuclear deterrent reflects the overarching and decidedly unhelpful air of secrecy in respect of all things nuclear. Yet it would not only be naïve to think that no progress has been made but would involve ignoring certain key elements in the discourse on India's nuclear deterrent which could prove instructive. It is to be noted that all of India's weapons delivery systems have been designed to be survivable – rail or road mobile – and relatively difficult to detect. From 2003, the doctrine has spoken of a command authority, and it makes explicit mention of alternate chains of command and measures to be taken to direct a retaliatory strike.

So, what has India achieved to date in respect of the criteria that have been outlined above for the credibility of its nuclear deterrent?

(1) Command and control

The closest that we have to an official statement in this respect comes from former National Security Advisor Shyam Saran, who delivered a lecture in 2013 on the credibility of the Indian nuclear deterrent. In respect of command and control, Saran made the following statement:

> The National Command Authority [NCA] is in charge of India's nuclear deterrent. At its apex is the Political Council which is headed by the Prime Minister and includes all the ministerial members of the Cabinet Committee on Security such as the Ministers of Defence, Home and External Affairs. Below the Political Council is the Executive Council which is headed by the National Security Advisor and includes the Chiefs of the three armed forces, the C-in-C of India's Strategic Forces Command, a three star officer, among others. There is an alternate National Command Authority which would take up the functions of nuclear command in case of any contingency when the established hierarchy is rendered dysfunctional. The NCA has access to radiation hardened and fully secured communications systems where, too, redundancies have been put in place as back-up facilities.[14]

This suggests that India has considered the command, control and communications difficulties that would cater for the possibilities of disruptions to the NCA in the event of an attack. Though he has not, in this part of his statement, indicated the survivability of the NCA, earlier on Saran made the following pertinent comments:

> India does have a credible theory of how its nuclear weapons may be used and that is spelt out in its nuclear doctrine. One may or may not agree with that doctrine but to claim that India does not have a credible theory about the use of nuclear weapons does not accord with facts. Since January 4, 2003, when India adopted its nuclear doctrine formally at a meeting of the Cabinet Committee on Security (CCS), it has moved to put in place, at a measured pace, a triad of land-based, air-delivered and submarine-based nuclear forces and delivery assets to conform to its declared doctrine of no-first use and retaliation only. It has had to create a command and control infrastructure that can survive a first strike and a fully secure communication system that is reliable and hardened against radiation or electronic interference. A number of redundancies have had to be created to strengthen survivability. India today has a long range ballistic missile capability and is on the road to a submarine-based missile capability. These capabilities will be further improved as time goes on and more

> resources become available. In all these respects, significant progress has been achieved. To expect that these should have emerged overnight after May 1998 is a rather naïve expectation. The record since the May 1998 nuclear tests demonstrates quite clearly a sustained and systematic drive to operationalize the various components of the nuclear deterrent in a manner best suited to India's security environment. This is not the record of a state which considers nuclear weapons as [an] 'instrument of national pride and propaganda'.[15] [author's emphasis]

Corroboration for these efforts at protecting the NCA has come from as far back as 2003, when the then government of Atal Bihari Vajpayee initiated the construction of two hardened shelters – reputedly capable of withstanding a direct hit from the equivalent of a B61 thermonuclear penetrating warhead – one of which has been established in Delhi, with an alternate within a 400km radius of the capital, enabling easy movement of the NCA in the event of need.[16]

One might assume that suitable command and control arrangements extend to ensuring launch orders are received by the various weapons crews as well as ensuring that the crews and their weapons are themselves protected in the event of attack. A 2010 *Pakistan Observer* report suggested that India was working on a more extensive shelter plan for military installations and using the metro lines in Indian cities to house shelters.[17] This secure command and control network is an essential part of making India's deterrent credible. It should be also noted that the survivability of the Indian NCA has not been discussed openly, nor has the nature of the alternate NCA.

Professor Bharat Karnad, writing in 2008, noted that the 2003 arrangements were somewhat makeshift, with one contingency involving a plan to move the prime minister via underground metro rail and thereafter by helicopter to a safe and secure location. Since then, with a scheduled completion in 2010/11, a dedicated nuclear command centre with two alternate sites has been designed and approved. These centres are designed to withstand the shockwave produced by a thermonuclear explosion as well as the blast. This would ensure the survivability and sustainability of the NCA.[18] Shyam Saran's statements in respect of the survivability of the NCA and its alternate may be indicative of the completion of this project.

India has made it clear that its NCA rests in civilian, political hands. This has given rise to the somewhat unwarranted comment that the Indian military is kept 'out of the loop' in respect of India's nuclear arsenal. This is a serious exaggeration and misrepresents the situation. Civilian control does not mean military exclusion. With the establishment of the Strategic Forces Command, the military has custody of the arsenal while control rests in political hands. Shyam Saran is particularly scathing in respect of this issue, saying:

> The thesis that India's nuclear deterrent is mostly symbolic is, for some, driven by the perception that India's armed forces are not fully part of the strategic decision-making process and that they play second fiddle to the civilian bureaucracy and the scientific establishment. Even if this perception was true, and in fact it is not, one cannot accept that the credibility of India's nuclear deterrence demands management by its military. The very nature of nuclear deterrence as practiced by a civilian democracy dictates that decisions relating to the nature and scope of the arsenal, its deployment and use, be anchored in the larger architecture of democratic governance.
>
> It is the civilian political leadership that must make judgments about domestic political, social and economic priorities as well as the imperatives imposed by a changing regional and global geopolitical environment. The military must be enabled to provide its own perspectives and inputs, just as other segments of the state must do. Undoubtedly the military's inputs and its advice would have to carry weight, especially in operational matters. But to equate exclusive military management of strategic forces, albeit under the political leadership's overall command, as the sine qua non of deterrence credibility is neither necessary nor desirable. One should certainly encourage better civil–military relations and coordination. It may also be argued that the military's inputs into strategic planning and execution should be enhanced to make India's nuclear deterrent more effective. But one should not equate shortcomings in these respects with the absence of a credible nuclear deterrent.
>
> In order to support the NCA, a Strategic Programme Staff has been created in the National Security Council Secretariat to carry out general staff work for the National Command Authority. This unit is charged with looking at the reliability and quality of our weapons and delivery systems, collat[ing] intelligence on other nuclear weapon states particularly those in the category of potential adversaries and work[ing] on a perspective plan for India's nuclear deterrent in accordance with a ten year cycle. The Strategic Programme Staff has representatives from the three services, from our Science and Technology establishment and other experts from related domains, including External Affairs. A Strategic Armament Safety Authority has been set up to review and to update storage and transfer procedures for nuclear armaments, including the submarine based component. It will be responsible for all matters relating to the safety and security of our nuclear and delivery assets at all locations. This will function under the direct authority of the NCA.[19]

This indicates that the services are part of the nuclear custody and command loop and, through the Strategic Programme Staff (SPS), has an important role in establishing a link between the political and military in respect of nuclear issues. While there is an argument that the entire defence decision-making process in India could benefit from more military input, it is manifestly not the case that the military is 'out' of the nuclear policy loop, and its perspective is considered by the NCA in the making of policy. The role of the recently created post of Chief of Defence Staff will be to act as military advisor to the NCA.

The SPS and the Strategic Forces Command (SFC) in themselves mark a departure from previously ad-hoc nuclear command and control systems. Professor Gaurav Kampani, one of the few writers dealing with practical aspects of the Indian SFC, wrote in respect of the SFC and its relationship to India's Integrated Defence Staff:

> However, this is not the case for India's nuclear forces, in which the SFC, although part of the IDS, has few linkages with it. The SFC commander reports directly to the chairman of the Chiefs of Staff Committee (CCoSC) and the NSA. Within the PMO [Prime Minister's Office], the SPS provides independent advice to the NSA. In effect, the command of India's nuclear forces flows from the PMO through the office of the NSA to the CCoSC and the SFC commander. More practically, the SFC commander interacts directly with the NSA, the SPS, and the NCA, more or less circumventing the conventional military entirely.[20]

In respect of the SFC itself, Kampani adds:

A Mirage 2000H upgraded to the latest Mirage 2000I standard. Note the MICA air-to-air missile under the outboard underwing pylon. (PIB)

> Since the middle of the last decade, the SFC's organizational presence within India's nuclear planning operation has grown substantially. In 2010, the former commander of the SFC, Rear Admiral (retired) Vijay Shankar, estimated the SFC's staff strength at a 'little below 100.' This number, according to Shankar, is a 'large staff for a command, especially when compared to other conventional operational commands, where staff strength does not exceed 50–60.' With organizational expansion, the SFC now has departments that cover logistics; a works department for building infrastructure; a technical section that has representation from all three military services; and a department of land, air, and sea vectors responsible for generating standard operating procedures for the various stages of operational readiness, in both peacetime and war. The SFC also has an electronics department that focuses on general release codes for nuclear weapons, as well as the NCA's general computing and communications requirements; an independent intelligence analysis group that processes raw data from various government agencies; and its own specialized medical staff.[21]

There are still questions regarding the manner of functioning of the NCA. For example, how will control work? It has been stated, by Saran among others, that the National Command Authority works on a two-person rule for access to armaments and delivery systems.[22] Does this include the use of permissive action links (PALs)? How will it function aboard a nuclear submarine or in the field with road and rail mobile systems? Would this require the deployment of DAE or BARC personnel alongside the military? The answers to these questions have yet to be disclosed publicly.

(2) Survivability and Credibility of the Arsenal

As discussed in detail in the relevant chapter, India's nuclear arsenal has taken the form of a nuclear triad with deployed road and rail mobile systems, a nascent or active airborne deterrent and developing submarine-based deterrent. In all these instances, the systems are designed to be mobile, hard to detect and as such survivable. As discussed in the aforementioned chapter, mountain tunnel complexes for the storage and protection of land-based missiles have been constructed, with more being planned.[23] It is yet unclear as to how many of these mountain complexes will be constructed, how secure they would be from satellite detection or how survivable they would be to a direct hit. Undoubtedly, they render the systems stored therein reasonably, even highly, survivable, but there are always doubts in this regard.

The Indian submarine-based deterrent is as yet in its infancy, with the INS *Arihant* still undergoing trials and its short-ranged K-15 missiles being impractical for deterrent purposes. Not until submerged launch tests of the K-4 missile are completed successfully will India's deterrent be deemed to have a viable triad. Even then, the 3,500km range of the K-4 is not particularly impressive and represents but a step towards a truly effective SSBN/SLBM (Ballistic Missile Submarine/Submarine-Launched Ballistic Missile) combination.

Saran also revealed that there are regular rehearsals for nuclear scenarios and escalatory situations, indicating that the Indian NCA and its supporting Strategic Programme Staff are catering for all contingencies:

> Regular drills are conducted to examine possible escalatory scenarios, surprise attack scenarios and the efficiency of our response systems under the no first use limitation. Thanks to such repeated and regular drills, the level of confidence in our nuclear deterrent has been strengthened. Specialized units have also been trained and deployed for operation in a nuclearized environment.[24]

Communicating Credibility

The Indian Strategic Forces Command rarely publicises any of its activities. The only time that its operations come to the fore are when it conducts user launches of ballistic missiles either as part of user trials or as routine training. The reticence of the SFC to publicise its operations and activities is somewhat unfortunate and not entirely desirable. This ties in to the one area where the credibility of the Indian deterrent suffers – visibility. By keeping virtually every aspect of the workings of the SFC, the NCA, the Strategic Programme Staff etc. secret, the Indian public is left largely, if not completely, unaware as to the capabilities and viability of the Indian deterrent. This is not satisfactory in an age where misinformation and rumour can undermine credibility more easily that one would like; it is necessary for a change in this posture.

In this respect, it is important to consider historian Satish Chandra's views in this regard:

> It is not sufficient to have a deterrence capability but also be perceived to have it as well as the will to use nuclear weapons if required to do so. In other words one must communicate or project the same to all concerned. Regrettably, insufficient attention has been paid to this aspect of establishing the credibility of our nuclear deterrent.[25]

India has taken steps to put some of these measures in place, but communication of these moves has been extremely poor – non-existent on occasion. Even Shyam Saran was moved to say:

> The secrecy which surrounds our nuclear programme, a legacy of the long years of developing and maintaining strategic capabilities, is now counter-productive. There is not enough data or information that flows from the guardians of our strategic assets to enable reasoned judgments and evaluations. There has been significant progress in the modernization and operationalisation of our strategic assets, but this is rarely and only anecdotally shared with the public. The result is an information vacuum which then gets occupied by either ill-informed or motivated speculation or assessments. To begin with, I would hope that the Government makes public its nuclear doctrine and releases data regularly on what steps have been taken and are being taken to put the requirements of the doctrine in place.
>
> It is not necessary to share operational details but an overall survey such as an annual Strategic Posture Review, should be shared with the citizens of the country who, after all, pay for the security which the deterrent is supposed to provide to them. An informed and vigorous debate based on accurate and factual information should be welcomed, because only through such debate can concepts be refined, contingencies identified and the most effective responses formulated. In a democracy, this is critical to upholding a broad consensus on dealing with the complex and constantly evolving security challenges our country confronts.[26]

While the actual number of weapons is unclear, as has been shown, India has the capability to make several hundred nuclear weapons and has retained the ability to surge weapons-grade plutonium production by keeping eight of its power reactors out of IAEA (International Atomic Energy Agency) safeguards. It is another matter as to whether India has in fact taken any such step to increase weapons-grade material production or to manufacture more weapons than its current inventory of delivery systems will permit. However, it is noteworthy – as indicated – that India has increased production of the Agni-I and Agni-II missiles and might be expected to do the same for the others in the Agni family in order to bolster its arsenal to the extent that it feels sufficiently numerically confident. As might be expected, India has done little by way of disclosing the extent of its missile production plans or its intended force levels.

India's response policies toward hostile actions by its neighbours, or hostile terrorist forces based on foreign soil, is undergoing a transformation, with cross-border ground and air strikes being launched in response to terrorist attacks from Pakistan in 2016 and 2019 and against terrorist forces in Myanmar in 2015. This was clearly part of an enhanced response policy adopted by an Indian government which no longer feels bound by threats of a nuclear response by Pakistan, and apparently marks a significant change in Indian security planning.

The only public photograph of an Indian nuclear weapon, possibly the Shakti-1 thermonuclear weapon during the 1998 tests. (PIB)

What has not been done is any articulation of these plans or policies in any coherent or even comprehensible way by official sources. This is patently absurd in the information age, and it would serve India well to devise a clear and comprehensive articulation of its nuclear position, its doctrine and its broad intent. To some extent, this was done through the draft nuclear doctrine, which did expound a rationale, an intent as well as a doctrine that was to inform India's development, deployment and use of nuclear weapons. However, as Dr Ali Ahmed pointed out:

> Here it must be acknowledged that the Draft was not to be taken as the government's position, even though it was released for discussion by the then National Security Advisor and Principal Secretary to the Prime Minister, Shri B Mishra. That it was not the official position was clarified after its release by a senior minister in the NDA [National Democratic Alliance] government, Shri Jaswant Singh. The Draft can, however, be taken as informing the doctrine officially adopted and explicated in the press release. There are elements of continuities and discontinuities between the two. However, the press release is the authoritative statement and is very clear.[27]

This means that analysts and adversaries are left to try and infer India's nuclear policy through two completely separate and somewhat different documents. That there was no official acceptance of the draft nuclear doctrine and the release of a mere summary of what was called India's nuclear doctrine adds to the unnecessary ambiguity, games of semantics and confusion surrounding Indian nuclear policy. A certain degree of ambiguity is not only desirable but also necessary. However, India's approach goes to the extreme and has resulted in an excessive level of confusion as to the country's interpretation of the wording of its own nuclear doctrine.

An Additional Area of Uncertainty – Non-traditional CBRN Attacks[28]

India faces not only the risk of traditional nuclear attacks, but also the potential use of nuclear weapons through non-traditional means. These could include a terrorist strike using a low-yield nuclear weapon, a so-called 'dirty bomb' or even an attack involving either a nuclear- or non-nuclear-generated electro-magnetic pulse. Another potential method of attack could be the possible sabotage of a nuclear power plant, with the potential danger of radioactive release. Strictly military or retaliatory measures will not necessarily deter such attacks. In this case, a combination of active measures to detect and prevent such attacks and passive measures to deal with their impact may have the best prospects of deterring or at the very least containing them.

The types of non-traditional CBRN (Chemical, Biological, Radiological, Nuclear) attacks may be broken down as follows:

- A nuclear or non-nuclear-generated electromagnetic pulse aimed at disrupting military and/or civilian activities, networks, operations and infrastructure
- the use of CBRN agents by a non-state actor aimed at creating either a mass casualty event or significant civilian fear and panic as well as disruption to society; or with the aim to stretch emergency, relief and economic resources with a view towards destabilisation of a state or to divert resources away from any military or diplomatic effort
- the sabotage of a chemical or nuclear power plant aimed at causing casualties and disruption through the release of radiation, radioactive materials or chemical agents

In none of the above scenarios do any of the conventionally accepted concepts of deterrence work effectively. Furthermore, with non-state actors as possible perpetrators, even the risk of retaliation may not prove sufficiently credible to deter such attacks. While efforts to prevent such non-traditional CBRN attacks would take the route of intelligence gathering, appropriate threat perception and prediction as well as the interception and/or interdiction of hostile entities intending to perpetrate these attacks, the ultimate guarantee of casualty minimisation as well as impact reduction lies in the Indian state being prepared to neutralise their effects. This can take the form of both proactive preventative measures and adequate disaster response mechanisms to deal with and minimise the deleterious effects of such attacks. Unfortunately, despite some significant progress, India's capabilities in this regard – which can be broadly described as civil defence measures – are as yet wholly inadequate and in need of improvement.

The principal agency tasked with dealing with planning for and thereafter mitigating the consequences of such an attack, the NDMA (National Disaster Management Authority), has been severely wanting in performance in the past. With new and professional personnel and a revamped structure, there is the possibility that it will perform somewhat more effectively in the future. Its efforts at

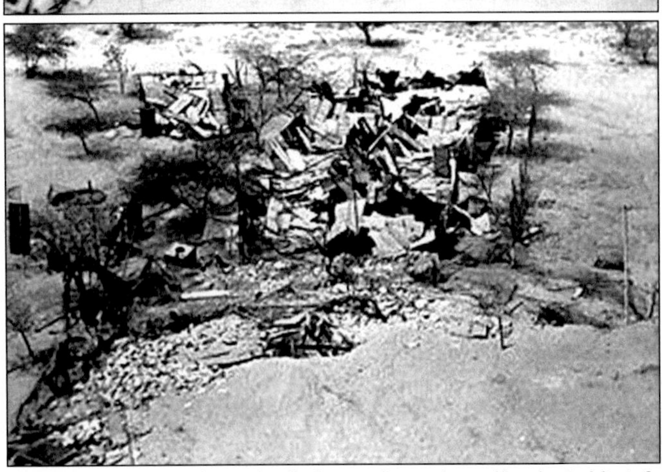

A series of photographs and stills from videos released by the Public Information Bureau of the Government of India, showing the site of the Shakti-1 thermonuclear test. (PIB)

Another series of photographs and stills from videos released by the PIB, this time showing the testing site of the Shakti-2 fission weapon. (PIB)

laying down guidelines and contingency plans for CBRN incidents are creditable, but more sustained efforts are needed to ensure effective measures.

The NDMA's functioning was reviewed after disasters in 2014 and has been significantly streamlined, with four expert members being appointed to its board and older, political appointees removed – the board being reduced from nine to four members. All four new members have extensive experience in disaster management and planning and should aid in the revitalisation of the NDMA. Furthermore, the rank equivalent of the members was downgraded to that of Secretary rather than the previously significantly higher rank of Minister of State. There has been a clear declaration of intent that the NDMA will no longer be a 'dumping ground' for politicians and bureaucrats with no experience, knowledge or expertise of any sphere related to disaster management.[29] It is to be noted that much of the practical experience of the NDMA has come from the handling of natural disasters. The agency has had no experience of handling mass casualty incidents caused by hostile action or chemical or nuclear accidents or incidents and has therefore drawn up plans based on best practices and simulations.

As the NDMA has the major coordination role in any disaster, communications have assumed importance, but their failure was seen to be responsible in part for a subsequent lack of a coordinated response. In recognition of that, a new mechanism for an uninterrupted communication network is being developed. The Department of Telecommunications has been tasked with the setting up of a robust communications system with a greater capacity to withstand natural disasters.[30] Command and control for the NDMA and the broader MHA is now facilitated by the National Emergency Operations Centre (NEOC), formerly known as the MHA Control Room, which the Indian draft national disaster management plan described as follows:

> An Emergency Operations Center (Control Room) exists in the nodal Ministry of Home Affairs, which functions round the clock, to assist the Central relief commissioner/designated officer in the discharge of his duties. The activities of the Control Room include collection and transmission of information concerning natural calamity and relief, keeping close contact with governments of the affected States, interaction with other Central Ministries/Departments/Organizations in connection with relief, maintaining records containing all relevant information relating to action points and contact points in Central Ministries etc., [and] keeping up-to-date details of all concerned officers at the Central and State levels.[31]

The cores of any response to a CBRN incident are the 12 battalions of the NDRF (National Disaster Response Force). These units, of which four are specialised CBRN formations, are dispersed geographically, and while their record has been very good thus

A still from a video showing the site of the Shakti-3 test. (PIB)

far, they have not been tested in a serious CBRN environment. Moreover, the decision not to equip all 12 battalions for CBRN as well as other disasters is seemingly irrational. While it might have been understandable in the early days of the NDRF, at present there exists no reason for the force not to be fully trained and equipped for such contingencies. This would enhance the flexibility of the NDRF substantially.

Unfortunately, the first responders to any CBRN incident will be India's beleaguered Fire and Emergency Services. These units are severely undermanned and underequipped. This continuing neglect is incomprehensible given the multi-hazard environment that exists in India, to which the contingency of a CBRN threat must be added. Equally neglected but even more poorly trained and equipped are the Home Guards and Civil Defence Volunteers, which are potentially invaluable organisations that could be of great assistance in the event of a disaster or an attack. Their continued neglect is unfortunate and should be reversed.

Finally, while India has some considerable capacity to cope with one or more localised non-traditional CBRN attacks, a sustained and large-scale attack – EMP (Electro-Magnetic Pulse), CBRN or multiple acts of sabotage – would present a qualitatively and quantitatively greater challenge. In such circumstances, it is unclear if India's existing systems would be able to provide an adequate response. It may be necessary, therefore, for India to review, reorganise and expand its existing systems to cope with a threat on a larger scale.

3
THE INDIAN ARSENAL TODAY[1]

India's nuclear arsenal is shrouded in excessive secrecy. While some of this is inevitable, the complete lack of information on the Indian nuclear weapons inventory, delivery options and command and control is somewhat disconcerting. However, the Indian arsenal – under the control of the Strategic Forces Command – is at present land-based and consists of air-delivered weapons and land-based ballistic missiles.

While India has made significant strides in developing a submarine-based deterrent, the INS *Arihant* and its K-4 are still a distance away from attaining operational status, though the recent two tests of the K-4 SLBM is noteworthy.

In the short to medium term, these systems will become relevant, but at this juncture, India would plan its nuclear deployment and delivery systems around manned aircraft and a mix of short-, medium- and long-range ballistic missiles. It is also to be noted that with the Nirbhay cruise missile facing development challenges and the BrahMos system's nuclear payload (if any) being in the realm of pure speculation, Indian nuclear armed cruise missiles cannot be

Unsafeguarded nuclear reactors at Chennai. (DAE)

An old view of the Dhruv nuclear reactor. (DAE)

counted on being part of the Indian arsenal at this juncture, but they are certainly prospectively so.

Given the fact that India has an avowed no first-use policy, this imposes a number of challenges for India to ensure the survivability of its deterrent. Having a nuclear triad is one way to ensure survivability, but their individual components must themselves be sufficiently survivable to conduct their operations in the event of a first strike on India. To this end, issues of numbers of delivery systems and warheads, the survivability of those assets in the event of a first strike and the ability of those delivery systems to reach their targets and deliver an effective response are all matters to be considered. While there is little dispute in India about the necessity for strategic and sub-strategic warheads, the issue of tactical nuclear warheads is as yet unsettled, with some notable analysts not being in favour of such systems.

India has made some progress towards developing a survivable and viable nuclear triad, with the individual components of manned aircraft, road- and rail-mobile IRBMs and ICBMs, submarine-launched ballistic missiles and land-based cruise missiles largely in place. However, development work in some respects is still ongoing regarding the submarine and cruise missile components of the Indian triad.

What is much less clear is whether adequate provisions have been made to build and deploy sufficient numbers of the delivery systems and to ensure that these systems are adequately protected in the event of a counter-force strike. This is not to say that India has not taken such steps, but the pervasive secrecy that permeates all aspects of the nuclear discourse in India makes certainty impossible.

The Fissile Material Stockpile

Fissile Material Stockpiles – Correlation to Nuclear Weapons Stocks[2]

India's fissile material stockpile is the subject of much debate. Some of this is borne out of the inevitable uncertainty surrounding the Indian nuclear weapons programme and its link to India's civilian reactor projects. However, a large and often contradictory debate has been raging involving groups that might charitably be termed as having 'vested interests' in either understating or exaggerating India's fissile material stockpile.

New Delhi's stoic silence in this debate has contributed to uncertainty, especially as concerns over Pakistan's purportedly growing fissile material stockpile has raised much consternation in the West. This has had the effect of decreasing scrutiny of the potential for Indian flexibility in the production of fissile material. While this is undoubtedly useful for Indian geopolitical objectives, for an analyst, it is unfortunate that India has done such a remarkable job at obfuscation.

Estimates of the Indian fissile material stockpile must consider that India has effectively three nuclear reactor programmes that are at once connected and separate. The first is the overtly military programme which once consisted of two reactors – CIRUS and Dhruva – before the former's decommissioning in 2010. To date, no reactor has been built to take its place, leaving Dhruva as the sole surviving specifically military reactor. In addition to these military reactors, there is India's large civilian reactor programme. Eight of these reactors have been kept out of international safeguard agreements and inevitably cause confusion in calculations of India's fissile material stockpile, leading to varying estimates.[3] Indeed, one has to ask whether India has not tasked at least one or two of its civilian unsafeguarded reactors to replace the decommissioned CIRUS. India is under no obligation to reveal any such move, nor is it one which will show any visible evidence of a change of use.[4]

The Indian nuclear industry – categorised as 'safeguarded civilian', 'unsafeguarded civilian' and 'military' – may be summarised as shown in Table 5.

Under the terms of the NSG (Nuclear Suppliers Group) Waiver granted to India following the India–US Civil Nuclear agreement, there are only two categories – safeguarded civilian and military. If India has chosen to make use of these facilities for its nuclear weapons programme, this inevitably causes all calculations on India's nuclear stockpile to go awry.

The Pakistani Nuclear Command Authority suggested that India had fissile stocks to make over 2,000 weapons,[6] a figure that can be traced back to a 2014 assessment by Mansoor Ahmed, a Pakistani nuclear analyst.

By Ahmed's calculations, he estimated that at the end of 2013, India's fissile material stockpile included 800–1,000kg of weapons-grade plutonium, 2 metric tons of highly enriched uranium (HEU) and 15 metric tons of reactor-grade plutonium.[7] Ahmed used the assumption that 4kg of weapons-grade plutonium, 50kg of HEU or 8kg of reactor-grade plutonium would be necessary to make one nuclear warhead, although the type or yield was not specified. It might be conceivable that Ahmed was using the calculations normally employed for a 20kt fission weapon (although his HEU calculations seem high for that). With these figures, he estimated that India could produce 250 warheads from weapons-grade plutonium, 40 from HEU (gun-type implosion devices, not thermonuclear weapons) and 1,875 from reactor-grade plutonium – for a total potential of 2,165 nuclear weapons.[8]

Ahmed's study further postulates that India is already working to install more than five fast breeder reactors which will increase

Table 5: Civilian Nuclear Facilities Under Continuous Safeguards[5]		
Type	**Facility, Location**	**Purpose, Design Capacity**
Thorium and uranium mines and mills	None – India's additional protocol does not apply to mines and mills	
Enrichment	None – India has not placed any enrichment facilities under safeguards	
Uranium conversion and fuel fabrication – natural uranium for PHWRs	Nuclear Fuel Complex, Hyderabad: Uranium Oxide Plant (Block A)	Conversion to UO2 (450t/year)
	Nuclear Fuel Complex, Hyderabad: Ceramic Fuel Fabrication Plant (Pelletising) (Block A)	Fuel fabrication (335t/year)
	Nuclear Fuel Complex, Hyderabad: Ceramic Fuel Fabrication Plant (Assembly) (Block A)	Fuel fabrication (300t/year)
	Nuclear Fuel Complex, Hyderabad: Gadolinia Facility	Zirconium alloy tubing and production
Uranium fuel fabrication – enriched uranium for light water power reactors	Nuclear Fuel Complex, Hyderabad: Enriched Uranium Oxide Plant and Enriched Fuel Fabrication Plant	Fuel fabrication using imported LEU (low enriched uranium) (24t/year)
Heavy water production	None – India's additional protocol does not apply to its heavy water production	
Power reactors – PHWRs	Rajasthan Atomic Power Station (RAPS1-6), Kota	PHWR, 90MW, 187MW, 202MWx4
	Kakrapar Atomic Power Station (KAPS1-2), Surat	PHWR, 202MWx2
	Narora Atomic Power Station (NAPS1-2), Narora	PHWR, 202MWx2
Power reactors – LWRs (Light Water Reactors)	Tarapur Atomic Power Station (TAPS1-2), Boisar	BWR, 150MWx2 (spent fuel to Tarapur storage)
	Kudankulam Nuclear Power Plant (KK1-2), Kudankulam	VVER-PWR, 917MWx2
Research reactors	None – India has not placed any research reactors under safeguards	
Fuel storage	Tarapur – Away from Reactor (AFR) Wet Spent Fuel Storage, Boisar	275t spent fuel
	Tarapur – Nuclear Power Plant site, Dry Spent Fuel Storage, Boisar	20t spent fuel
	Other safeguarded spent fuel from PHWRs in storage	About 1,500t spent fuel
Reprocessing and separated plutonium	Option to construct safeguarded reprocessing plants for PHWR fuel under US–India Nuclear Cooperation Agreement	
	Approximately 0.4t of separated plutonium currently under safeguards, having been reprocessed at PREFRE-1, while the facility was under temporary safeguards	
Other research centres	None – India's additional protocol does not apply to sites where India conducts nuclear fuel cycle-related research and development	

its weapons-grade plutonium production capacity by 20 times to 700kg every year. In addition, he suggested an expansion in India's centrifuge enrichment programme will enable it to increase production of highly enriched uranium for nuclear weapons to 160kg every year. Should these calculations be accurate, India could theoretically produce approximately 80–90 plutonium-based weapons and seven or eight uranium-based nuclear weapons annually.[9]

The *Bulletin of the Atomic Scientists*, in a response, suggested a more realistic upper estimate of 844 weapon equivalents if reactor-grade plutonium is considered, falling to a generous estimate of 219 if only the CIRUS and Dhrvua outputs are considered, with a more likely figure of 75–125 weapons being likely, which possibly coincides with India's missile arsenal to date.[10]

In support of their contentions, the *Bulletin of the Atomic Scientists* suggested that, based on a 2015 study by the Institute for Science and International Security (ISIS) at the end of 2014, India likely possessed about 550kg of weapons-grade plutonium, 100–200kg of HEU intended for use in thermonuclear weapons and 2.9 metric tons of separated reactor-grade plutonium. The ISIS study concluded that this fissile material was sufficient to produce between 75 and 125 nuclear warheads. ISIS arrived at this number mainly through an appraisal of India's probable weapons-grade plutonium stockpile, assuming that India would not use reactor-grade plutonium in nuclear warheads, and that HEU would only

Table 6: Facilities Used Primarily for Civilian Purposes and not Subject to Continuous Safeguards

Type	Facility, Location	Purpose, Design Capacity
Thorium and uranium mines and mills	See Table 7 on military facilities below – in addition to being a source of nuclear material for India's military programme, domestic mines are an important source for several reactors	
Enrichment	See Table 7 on military facilities – Rare Materials Plant (RMP) has both military and civilian applications	
Uranium conversion and fuel fabrication	Some facilities in Nuclear Fuel Complex, Hyderabad	For unsafeguarded PHWRs
	Plan for second Nuclear Fuel Complex, Kota	
	Trombay Fuel Fabrication	Small-scale production for Fast Breeder Test Reactor
Heavy water production – declared civilian as part of Separation Plan	Hazira	80t/year
	Thal-Vaishet	78t/year
	Tuticorin	49t/year
Power reactors – PHWRs	Tarapur Atomic Power Station (TAPS3-4), Boisar	PHWR, 490MWx2
	Madras Atomic Power Station (MAPS1-2), Madras	PHWR, 202MWx2
	Kaiga (KGS1-4), Kaiga	PHWR, 202MWx4
Power reactors – FBRs	Prototype Fast Breeder Reactor, Kalpakkam	FBR, 500MW (scheduled for April 2016)
Power reactors – other	Advanced Heavy Water Reactor	HWR, 300MW (not yet operational)
Research reactors	Fast Breeder Test Reactor, Kalpakkam	FBR, 40MWt
	Purnima reactor, Trombay	Small-scale U-233 LWR
	Kamini reactor, Kalpakkam	U-233-fuelled LWR, 30kWt
	Apsara reactor, Mumbai	LWR, 1MWt (undergoing conversion from HEU fuel to LEU)
	Compact High Temperature Reactor, Trombay	U-233 and Th fuel (under construction)
Fuel storage	Rajasthan NPP site, Away From Reactor (AFR) Dry Spent Fuel Storage, Rajasthan	570t spent fuel
	Spent fuel storage for unsafeguarded PHWRs to be reprocessed	2,500–3,600t spent fuel
Reprocessing and separated plutonium	Kalpakkam Spent Fuel Reprocessing (KARP)	PUREX (used for both military purposes and reuse in civilian reactors), 100t/year spent fuel
	Tarapur Power Reactor Fuel Reprocessing (PREFRE-1 and 2) Centre	PUREX (used to reprocess spent fuel from PHWRs for reuse in civilian reactors; temporary safeguards apply on 'campaign' basis), 100–150t/year spent fuel
	CORAL reprocessing plant, Kalpakkam	Pilot plant for reprocessing spent fuel from FBRs, 12kg/year
	Tarapur Advanced Fuel Fabrication Facility	MOX fuel fabrication for various reactors

Table 7: Facilities Primarily Used for Military Purposes		
Type	**Facility, Location**	**Purpose, Design Capacity**
Thorium and uranium mines and mills	Jaduguda Mill	200t/year
	Turamdih Mill	190t/year
	Tummalapalle Mill	220t/year
	More mills to enter operation in the next few years	
	Mines in Jharkland, Andhra Pradesh, Telengana, Karnataka, Meghalaya	
Enrichment	Rare Materials Plant (RMP), Bhabha Atomic Research Center, Mysore/Rattehalli (Karnataka)	Gas centrifuge enrichment for naval reactor fuel, LEU for Apsara reactor, possibly also for nuclear weapons, 15–25t SWU*/year (capacity to produce 60–100kg of weapons-grade uranium/year from natural uranium feed)
	Apparent second RMP under construction at same site	Larger than original RMP
	Special Material Enrichment Facility (SMEF), Khudapura (Karnataka), in initial stages of construction	Gas centrifuge enrichment for civilian and military purposes, larger than RMP
	Uranium Enrichment Plant, Trombay	Pilot-scale research on ultracentrifuges
	Laser enrichment research, various locations	
Uranium conversion and fuel fabrication – natural uranium	Uranium metals plant, Trombay	Fuel fabrication for CIRUS and Dhruva reactors
Heavy water production – not declared civilian as part of 'Separation Plan'	Baroda	17t/year
	Kota	85t/year
	Manuguru	185t/year
	Talcher	62t/year
Research reactors	Dhruva, Mumbai	HWR, 100MWt (historically a major source of weapons-grade plutonium for nuclear weapons)
	CIRUS, Mumbai	HWR, 40MWt (historically a major source of weapons-grade plutonium, despite being supplied by Canada in the 1950s on condition that it only be used for peaceful purposes; shut down in 2010 as part of India's Separation Plan)
Naval reactors	Advanced Technology Vessel naval prototype PWR (HEU fuel), Kalpakkam	80–100MWt each
	INS *Arihant* SSBN with PWR (HEU fuel)	
	Two similar SSBNs under construction, two more planned	
Reprocessing	Trombay plutonium separation plant, Bhabha Atomic Research Centre	400–700kg weapons-grade plutonium separated as of 2014. PUREX, approximately 50t spent fuel per year, including reprocessing weapons-grade plutonium from Dhruva, primarily for nuclear weapons
* SWU = Separative Work Unit, the standard measure of the effort required to separate uranium isotopes U235 and U238		

be used to produce a handful of thermonuclear weapons at most. ISIS also considered that some plutonium is in weapons production pipelines or held in reserve, assessing that only about 70 percent of India's stockpile is available to be made into weapons and assumed that it would take 3–5kg of weapons-grade plutonium for each warhead.[11]

The biggest difference between these two estimates comes from their assessments of, and assumptions about, reactor-grade plutonium. The ISIS study completely discounts the possibility that India would use reactor-grade plutonium in its nuclear weapons. Furthermore, its estimate of India's reactor-grade plutonium stockpile is also significantly lower than Ahmed's: 2.9 metric tons as opposed to 15 metric tons.

India's reprocessing plants have not had a particularly impressive record to date, although there is some evidence to suggest that Western estimates of their poor performance are greatly overstated.[12] This, unfortunately, is a potential pitfall in using the estimates and calculations of India's capabilities.

The *Bulletin of the Atomic Scientists* suggests that, discounting reactor-grade plutonium entirely, India could possibly build some 219 weapons.[13] It further asserts that 'much, or even most, of India's HEU is intended for use in naval reactors rather than in nuclear warheads'.[14] It is also clear that some of India's weapons-grade plutonium was already used in nuclear tests or is contained in process waste, which further reduces the estimate to between 75 and 125 weapons.

Table 8: Indian and Pakistani Fissile Material Production Capacity[18] *– Estimates (Existing and Projected)*

	India	Pakistan
Production Reactors	100 MWth* Existing 155 MWth Planned	160–200 MWth
Breeder Reactors	500–600 MWe Existing 1,200 MWe Planned	Nil
PHWRs	2,350 MWe Existing 2,800 MWe Planned[252]	Nil
Reprocessing Plants	350 tHM**/year Existing 1650 tHM/year Planned	140 tHM/year Existing
Uranium Enrichment	30,000–45,000 SWU/year Existing 126,000 (Planned)	15,000–45,000 SWU/year
Fissile Material Stockpiles	3.1 tons HEU (1.1 tons U-235) 0.7 tons WG Pu*** 5.5 tons RG Pu (separated) 11–14 tons RG Pu (unseparated)	3.2 tons HEU 0.19 tons WG Pu No RG Pu

* MWth = megawatts of thermal output
** tHM = tonnes of heavy metal
*** WG Pu = weapons-grade plutonium; RG Pu = reactor-grade plutonium

Table 9: Indian and Pakistani Fissile Material Weapon Equivalent Potential

	India	Pakistan
WG Pu	0.59–0.79 tons	0.19–0.21 tons
RG Pu	5.5 tons (Separated) 11–14 tons (Unseparated)	0 tons
HEU	3.2 tons (1 ton U-235)	3.1 tons
Weapons Potential	148–198 (WG Pu) 688 (RG Pu Separated) 1,375–1,759 (RG Pu Unseparated) 50 (HEU) = 2,261–2,686	48–52 (WG Pu) 0 (RG Pu) 155 (WG HEU) = 207

It is a legitimate question to ask why India would wish to use reactor-grade plutonium for the production of nuclear weapons. Reactor-grade plutonium, with its higher content of the Pu-240 isotope, could produce pre-initiation in weapons so produced. However, reliable weapons with yields from low kiloton range up to 20–40kt with reactor-grade plutonium – comprising a fissile content of 66 percent – requires a slightly higher critical mass than weapons-grade plutonium.[15] Furthermore, pre-initiation problems could be dealt with through using the deuterium-tritium 'boosting' of fission devices.[16] The use of such D-T 'boosting' enables the manufacture of reliable and efficient weapons using reactor-grade plutonium, with the 'boosting' rendering such weapons virtually immune from pre-initiation problems. It would therefore mean that discounting India's stocks of reactor-grade plutonium is not necessarily wise when examining India's nuclear weapons potential.

It is therefore not feasible to ignore India's civilian unsafeguarded reactors and their potential. Mansoor Ahmed, writing for the Belfer Centre, after the *Bulletin of the Atomic Scientists* response, summarises current and future Indian and Pakistani fissile material stocks and weapons equivalents as follows in tables 8 and 9:[17]

From these tables, we can see that there is a wide divergence in respect of India's fissile material stockpile. Whether India has made any use of its civilian facilities for weapons-grade plutonium production is a matter of conjecture. This would inevitably lead to an exponentially higher demand for uranium for such reactors, but such costs would be offset by eliminating the need to build and operate additional military reactors. This approach has evoked much speculation, as analysts such as Bharat Karnad and, during the 1980s and 1990s, even non-proliferation advocates such as Leonard Spector have suggested that India has pursued this route.

Using PHWRs in the 'low burn-up' mode would inevitably require them to be refuelled more often, thus increasing the quantity of uranium for fuelling the reactors. To give an indication as to the quantitative difference in the uranium required, Dr Ashley Tellis, writing in support of the India–US civil nuclear agreement at a time when concerns in this regard were raised, suggested the figures shown in tables 10.1–10.4, comparing fuel requirements for the dedicated military reactors and the civilian reactors, and then examining the requirements for civilian reactors run for military purposes.[19]

The unfortunate corollary to these conflicting figures means that estimates of Indian fissile material stocks and their disaggregation into civilian, military, weapons-grade and reactor-grade is largely a matter of conjecture.

This is further complicated by the fact that India's CANDU (Canadian Deuterium Uranium) PHWRs typically operate at relatively lower burn-ups of 6,700 Mega-Watt days per ton (MWd/t) compared to Light Water Reactors (LWRs), which run on higher

TABLE 1. ***Estimated Amounts of Uranium Required for Producing Weapons-Grade Plutonium at Various Capacity Factors in India's Research Reactors***

Reactor	Capacity Factor (%)	Thermal Power (MW)	Thermal Energy Output (MWD/yr)	Average Discharge Burnup (MWD/MTU)	Fuel Requirement (MTU/yr)	WGPu per year (kg)[a]
CIRUS						
	40	40	5,840	665	8.8	5.4
	40	40	5,840	1,000	5.8	5.1
	40	40	5,840	1,400	4.2	4.9
	50	40	7,300	665	11.0	6.7
	50	40	7,300	1,000	7.3	6.4
	50	40	7,300	1,400	5.2	6.1
	70	40	10,220	665	15.4	9.4
	70	40	10,220	1,000	10.2	9.0
	70	40	10,220	1,400	7.3	8.7
Dhruva						
	65	100	23,725	665	35.7	21.8
	65	100	23,725	1,000	23.7	20.9
	65	100	23,725	1,400	16.9	19.9
	75	100	27,375	665	41.2	25.2
	75	100	27,375	1,000	27.4	24.1
	75	100	27,375	1,400	19.6	23.0

[a] Pu production is calculated at 1 kg/1000 MWD of thermal energy output multiplied by a correction factor based on burnup level.

Notes. MW, megawatt. MWD/yr, megawatt days per year. MWD/MTU, megawatt days per metric ton of uranium. MTU/yr, metric ton of uranium per year. WGPu, weapons-grade plutonium. Kg, kilogram.

Table 10.1 (Tellis)

TABLE 2. ***Estimated Fuel Requirements for Operational Pressurized Heavy Water Reactors***

Reactor	MWe	Thermal Power (MW)	Annual Thermal Energy Output (MWD/yr)[a]	Annual Fuel Requirement (MTU/yr)	Date of Commercial Operation	Notional Decommissioning Date[b]	Fuel Needed 2006–Decommissioning (MTU)
RAPS-1 Rawatbhata, Rajasthan[c]	100	350	93,258	0.00	December 1973	?	0.00
RAPS-2 Rawatbhata, Rajasthan	200	690	183,759	27.43	April 1981	2011	137.13
RAPS-3 Rawatbhata, Rajasthan	220	760	202,502	30.22	June 2000	2030	725.38
RAPS-4 Rawatbhata, Rajasthan	220	760	202,502	30.22	December 2000	2030	725.38
MAPS-1 Kalpakkam, Tamilnadu	220	760	202,502	30.22	January 1984	2014	241.79
MAPS-2 Kalpakkam, Tamilnadu	220	760	202,502	30.22	March 1986	2016	302.24
NAPS-1 Narora, Uttar Pradesh	220	760	202,502	30.22	January 1991	2021	453.36
NAPS-2 Narora, Uttar Pradesh	220	760	202,502	30.22	July 1992	2022	483.59
KAPS-1 Kakrapar, Gujarat	220	760	202,502	30.22	May 1993	2023	513.81
KAPS-2 Kakrapar, Gujarat	220	760	202,502	30.22	September 1995	2025	574.26
KAIGA-1 Kaiga, Karnataka	220	760	202,502	30.22	November 2000	2030	725.38
KAIGA-2 Kaiga, Karnataka	220	760	202,502	30.22	March 2000	2030	725.38
TAPS-3 Tarapur, Maharashtra[d]	540	1,860	495,597	73.97	May 2006	2036	2,219.09
TAPS-4 Tarapur, Maharashtra	540	1,860	495,597	73.97	September 2005	2035	2,145.12
Total				**478**			**9,972**

[a] Assumes a capacity factor of 73 percent.

[b] Assumes 30-year reactor life.

[c] Reactor virtually non-operational.

[d] TAPS-3 went critical in May 2006, but is treated as operational for calculation purposes in this report.

Notes. Reactors identified in shaded rows have been offered for safeguards in India's March 2006 separation plan. MWe, megawatt-electric. MW, megawatt. MWD/yr, megawatt days per year. MTU/yr, metric ton of uranium per year. MTU, metric ton of uranium.

Table 10.2 (Tellis)

fuel burn-ups of 33,000–50,000 MWd/t.[20] India's PHWRs, operating at normal capacity factors and an average burn-up of 6,700 MWd/t, would therefore produce less of the more dangerous plutonium isotopes such as Pu-240 compared to LWRs, thus making the plutonium so produced more readily useable for nuclear weapons production.[21]

TABLE 3. ***Estimated Fuel Requirements for Pressurized Heavy Water Reactors Currently under Construction***

Reactor	MWe	Thermal Power (MW)	Annual Thermal Energy Output (MWD/yr)[a]	Annual Fuel Requirement (MTU/yr)	Scheduled Date of Commercial Operation	Notional Decommissioning Date[b]	Total Fuel Needed Until Decommissioning (MTU)[c]
RAPS-5 Rawatbhata, Rajasthan	220	760	202,502	30.22	2007	2037	932.50
RAPS-6 Rawatbhata, Rajasthan	220	760	202,502	30.22	2008	2038	932.50
KAIGA-3 Kaiga, Karnataka	220	760	202,502	30.22	2007	2037	932.50
KAIGA-4 Kaiga, Karnataka	220	760	202,502	30.22	2007	2037	932.50
Total				**121**			**3,730**

[a] Assumes a 73 percent capacity factor.

[b] Assumes 30-year reactor life.

[c] Includes an initial loading of 56 MTU.

Notes. Reactors identified in shaded rows have been offered for safeguards in India's March 2006 separation plan. MWe, megawatt-electric. MW, megawatt. MWD/yr, megawatt days per year. MTU/yr, metric ton of uranium per year. MTU, metric ton of uranium.

Table 10.3 (Tellis)

TABLE 4. ***Comparison of Notional 220 MWe Reactor Fuel Requirements in Production of Weapons-Grade Plutonium versus Electricity***

Capacity Factor (%)	Thermal Power (MW)	Annual Thermal Energy Output (MWD/yr)	Average Discharge Burnup (MWD/MTU)	Fuel Requirement (MTU/yr)	WGPu per year (kg)[a]
Weapons-Grade Plutonium					
73	760	202,502	665	304.51	186.30
73	760	202,502	1,000	202.50	178.20
73	760	202,502	1,400	144.64	170.10
79	760	219,146	665	329.54	201.61
79	760	219,146	1,000	219.15	192.85
79	760	219,146	1,400	156.53	184.08
Electricity					
Capacity Factor (%)	Thermal Power (MW)	Annual Thermal Energy Output (MWD/yr)	Average Discharge Burnup (MWD/MTU)	Fuel Requirement (MTU/yr)	WGPu per year (kg)
73	760	202,502	6,700	30.22	NA
79	760	219,146	6,700	32.71	NA

[a] Plutonium production is calculated at 1 kg/1,000 MWD of thermal energy output multiplied by a correction factor based on burnup level.

Notes. MWe, megawatt-electric. WGPu, weapons-grade plutonium. MW, megawatt. MWD/yr, megawatt days per year. MWD/MTU, megawatt days per metric ton of uranium. MTU/yr, metric ton of uranium per year. Kg, kilogram. NA, not applicable.

Table 10.4 (Tellis)

India need not dedicate all eight of its unsafeguarded reactors to fissile material production. Given the fact that the burden of India's weapons-grade plutonium production is now borne entirely by the Dhruva research reactor, allocating one or two reactors to fissile material production would significantly boost India's production of such material. Tellis suggests that one reactor – even a small 220MWe reactor – could be so tasked without becoming excessively costly.[22] This could potentially add 150–200kg of weapons-grade plutonium per year to India's stockpile.[23]

To further complicate calculations, Professor Karnad asserts that from 2001, India had operated some of its unsafeguarded PHWRs in 'low burn-up' mode to bolster its fissile material stocks, to the extent that by 2010–2012, sufficient fissile material was on hand for 200 nuclear weapons.[24] He further states that plans suggest India desires a force of some 300–400 weapons by 2030.[25] With the potential use of the PHWRs to produce weapons-grade plutonium or by using reactor-grade plutonium with deuterium-tritium 'boosting', India may well be on the way to reaching that target.

India's Warheads[26]

Just as India's delivery vehicles can be divided into the manned aircraft and ballistic/cruise missile categories, India's warheads must also be split into those intended to be low-yield weapons for use against tactical targets and larger-yield strategic systems.

Twin domes of the unsafeguarded reactors at Tarapur. (PIB)

To discount these two statements coming from two totally different perspectives – one by a former Chairman of the Chiefs of Staff Committee and the other by the former Director-General of the DRDO – would be folly to say the least and could lead any analysis of India's nuclear arsenal to fall into serious error. The Indian design team has admitted that the trigger in the Shakti-1 thermonuclear device tested in 1998 was a fusion-boosted-fission device.[34] It should be noted that even a recognised sceptic like Professor Bharat Karnad remarks that India's boosted-fission capability is significantly more reliable than its thermonuclear capability.[35]

From the data gleaned from the French fission and boosted-fission designs, it is clear that the weight of such weapons at even higher yields fits in easily with India's Agni family of missiles, which have payloads ranging from 1,000kg to 1,500kg. Professor Karnad asserts that the Agni-I has been optimised for a 20–30kt warhead, the Agni-II for a 90–150kt warhead and the Agni-III for a 300kt warhead.[36] However, issue should be taken with the first named missile, as a 20–30kt warhead ought not to approach 1,000kg (which is the payload of the Agni-I), which runs counter to the reduced weight of both the 15–20kt and 100kt fission weapons developed in the 1980s.[37] Rather, it is likely that the Agni-I has the same warhead as the Agni-II, being deployed with a warhead in the 150kt range.

India's nuclear warheads are the subject of intense and often contradictory speculation, with experts such as Dr Tellis being vociferous in suggesting only 15–20kt warheads are deployed and deemed reliable. However, Professor Bharat Karnad, in some of his comprehensive work on the subject, notes that 100kt fission weapons had been produced as early as the 1980s.[27] This would suggest that weapons in the ranges of 15–100kt would be the main service types for the Indian nuclear arsenal, whether air- or missile-based.[28] The fact that the Indian government has officially only indicated that yields of 200kt are deployable should not be doubted, especially considering the probability that 100kt fission weapons being in existence.[29]

There is some anecdotal evidence for India developing fission weapons approaching these French weapons in yield. Indications are that the first Indian nuclear weapons design had a mass of about 1,000kg, with a yield of 12–15kt. However, subsequently, perhaps by 1982 (when rumours of a fresh round of nuclear tests were being circulated), the said weapon had been scaled down to a more manageable mass of between 170kg and 200kg.[30] Later, it appears that a 100kt fission weapon was produced for aerial delivery with a mass of 200–300kg.[31] If this is accurate, it would mean that India had perfected a relatively high-yield fission weapon with a relatively low mass for its class. One would expect that missile warheads of similar designs and yields would be feasible.

That India might have fielded such fusion-boosted-fission weapons is borne out in two interesting sets of statements. Firstly, Admiral Arun Prakash in 2009 (at the height of the controversy generated by Dr Santhanam's statements on the 1998 tests) said: 'In the midst of the current brouhaha, we need to retain clarity on one issue; given that deuterium tritium boosted-fission weapons can generate yields of 200–500 kt, the credibility of India's nuclear deterrent is not in the slightest doubt.'[32]

More recently, an even more interesting comment was made by Dr Avinash Chander in 2011 to the *Business Standard* newspaper: 'Now we talk of [accuracy of] a few hundred metres. That allows a smaller warhead, perhaps 150–250 kilotons, to cause substantial damage.'[33]

Indian Thermonuclear Weapons?

The question as to whether India has any deployed thermonuclear weapons is one that inevitably brings the disputed yield of the Shakti-1 thermonuclear test to the fore. This has sharply divided opinions among analysts, with allegations, rebuttals and unfortunate forays into character assassination becoming part of the discussion. Rather more unfortunate is the tendency of aggressive Western advocates of containing or capping the Indian nuclear weapons programme to use the disputed thermonuclear yield to ridicule India's capability while simultaneously pushing for more aggressive action aimed at coercing India.

If some Indian analysts are to be believed, thermonuclear weapons with yields of between 125kt and 175kt are readied for deployment.[38] However, there is little doubt that the controversy over the success of the Shakti-1 test would cause the user to have doubts as to the efficacy of these designs. This is compounded by the fact that unlike fission or even boosted-fission weapons, the physics involved in the manufacture and development of thermonuclear weapons does not lend itself to single-test perfection of an entire scalable range of weapons design. This would inevitably mean that for India's Strategic Forces Command to be confident in the yield of these weapons, a resumption of nuclear testing would be required. The corollary to this would be that India might not be sufficiently confident to deploy such weapons.

However, Dr Anil Kakodkar, former Chairman of the Atomic Energy Commission, in an interview with Karan Thapar of CNN-IBN, categorically stated that India had thermonuclear weapons. The text of part of that interview is reproduced below:

Karan Thapar: 'Let me put to you two or three critical issues. Given the fact that you have concluded several reviews, including one recently after the doubts were raised, the doubts continue. And given that there are doubts about India's one and only thermonuclear test do we need more tests?'
Anil Kakodkar: 'Well, I would say no because the important point to note is that the thermo nuclear test, the fission test and the sub-kilotonne test all worked as designed. They are diverse.

In terms of detailed design, their content is quite different. And so we think that the design which has been done is validated and within this configuration which has been tested one can build devices ranging from low kilotonne all the way to 200 kilotonnes. And that kind of fully assures the deterrence.'
Karan Thapar: 'You are saying that India doesn't need more thermonuclear tests but the truth is that all the established thermonuclear powers needed more than one test. Can India be the exception?'
Anil Kakodkar: 'Well if you go by *Dil Maange More* ['the heart desires more'], that's another story. But we are talking about a time where the knowledge base has expanded, the capability has expanded and you carry out a design and prove you are confident that on the basis of that design and that test, one can build a range of systems right up to 200 kilotonnes.'
Karan Thapar: 'I want to pick up on that last point that you have just made. Given that doubts continue and given that there are going to be no further tests and you are not saying that there is any need for further tests – can you say India has a credible thermonuclear bomb?'
Anil Kakodkar: 'Of course.'
Karan Thapar: 'We have a credible thermonuclear bomb?'
Anil Kakodkar: 'Why are you using singular? Make that plural.'
Karan Thapar: 'The reason I ask is because Dr Santhanam writing in *The Hindu* says that the thermonuclear device has not been weaponsied even 11 years after the tests.'
Anil Kakodkar: 'How does he know? He is not involved.'
Karan Thapar: 'So you are saying to me that we have thermonuclear bombs – in the plural?'
Anil Kakodkar: 'Yes.'
Karan Thapar: 'With a yield of at least 45 kilotonnes each.'
Anil Kakodkar: 'Much more than that.'
Karan Thapar: 'Much more than that?'
Anil Kakodkar: 'Yes. I told you we have the possibility of a deterrence of low kilotonne to 200 kilotonnes.'
Karan Thapar: 'So when people like former Army chief, General Malik, say that because of the doubts in the public arena, the Army wants assurance of the yield and the efficacy of India's thermonuclear bomb, what is your answer to them?'
Anil Kakodkar: 'I think that is guaranteed. The Army should be fully confident and defend the country. There is no issue about the arsenal at their command.'[39]

In effect, Dr Kakodkar was indicating that the thermonuclear design had been weaponised and that the user could have confidence in its reliability. The question is whether these assurances are sufficient or the relationship between the warhead designer, the delivery system designer and the user sufficiently integrated to give the latter faith in those assurances. In this regard, an interesting anecdote is narrated by Professor Karnad to indicate that the relationship between the warhead designers and the missile designers has improved to the point where, over a decade ago, S.K. Sikka, who was head of the thermonuclear weapons project, was asked to reduce the diameter of a weapon intended to be fitted on an Agni variant by 5cm. His ready compliance with the request was deemed indicative of an improved relationship between design teams.[40]

Despite the credibility of the Indian deterrent being unaffected by a compromised thermonuclear capability, the development of thermonuclear weapons is an essential part of Indian weapons development, not to ensure the credibility of the deterrent but to make it more flexible. Thermonuclear weapons need not have higher yields than either fission or boosted-fission weapons – the French TN-75 fitted to the M45 SLBM having a yield of 100kt – but they are inevitably lighter, with the 300kt TN-80/81 warhead of the ASMP missile weighing a mere 200kg.[41]

With thermonuclear weapons relying on fusion for the majority of their yield, the primaries of fusion weapons can often use smaller quantities of fissile materials. Moreover, if, as India has done, a fusion-boosted-fission primary is used, then the possibility of a variable yield weapon can be considered. Indeed, shortly after the 1998 tests, Dr Frank Barnaby, writing in *Jane's Defence Weekly*, suggested that an operational nuclear weapon could have variable yields of 5kt, 50kt and 500kt.[42] Such flexibility obviates the need for India to maintain a separate inventory of fission weapons to provide lower-yield options alongside the larger fusion-boosted-fission weapons.

India's Warheads

To date, some 20 years after the 1998 nuclear tests, India has shown no inclination to resume nuclear testing. With this limitation, what are India's warhead options?

Large fission weapons – in excess of 50kt and up to, as Professor Karnad suggests, 100kt – would perhaps involve the use of too much fissile material. While India's fission design capabilities are easily its most reliable, its fissile stocks might mitigate against opting for this route. Rather, there seems to be little doubt that the boosted-fission primary of the Shakti-1 test worked satisfactorily. In light of this, it is likely that India's larger-yield weapons would be of the boosted-fission type. As noted above, both Admiral Prakash and Dr Chander suggest weapons with yields in the 150–250kt range. It is also known that the payload capacity of the Agni series of missiles hovers between 1,000kg and 1,500kg. This relatively large payload for a monolith warhead would perhaps suggest that these missiles are indeed fitted with boosted-fission warheads. While it is true that boosted-fission warheads can achieve yields of 500kt, the Indian nuclear establishment has claimed only yields of 200kt since 1998, thus perhaps indicating an upper limit on the yield of India's deployed boosted-fission arsenal though not precluding the development and deployment of higher-yield designs.

India has also examined the possibility of deploying Multiple Independently Targetable Re-entry Vehicles (MIRVs) on later variants of the Agni family. A new missile – the Agni-6 – will apparently test India's first MIRVs. However, deploying MIRVs will raise additional issues – namely, whether or not the number of launchers is adequate to ensure survivability. Insufficient numbers of launchers and missiles would lead to a disproportionate loss of warheads if MIRVs are deployed without a commensurate increase in the number of deployed missiles.

The choice for India is to find the correct compromise between the advantages offered by MIRV missiles and the requirement for larger numbers of missiles – albeit with single warheads. Whatever choice is made, India should increase the number of deployed missiles for credibility and survivability. MIRVs will place additional strain on the stocks of India's fissile materials, and as such, the deployment of MIRVs will require India to increase its stocks of fissile materials and its production of warheads. This may be part of the reason why India may be seeking thermonuclear weapons as, though there is no reason why fission or fusion-boosted-fission weapons cannot be deployed as MIRVs, the yield of the fission weapons thus employed would be limited; and as fusion-boosted-fission weapons would inevitably be much more bulky for their given yield than thermonuclear weapons, it is debatable whether it is practical to use them in MIRVs.

Are India's Warheads Mated or De-Mated to their Delivery Systems?[43]

For nearly a decade, the debate over whether India maintains a stock of fully mated nuclear warheads has been raging. Some analysts have suggested India's deterrent is 'recessed'. While this may have been true in the past, India is reaching the stage where a deployed force of warheads mated to missiles is becoming increasingly likely. In 2003, a BBC report suggested that India would take between eight and 12 hours to mate its warheads and retaliate to a first strike.[44] However, since then, with the creation of the Strategic Forces Command and a more streamlined process, it is probable that India does not need such a long period of time. Professor Karnad suggested that, in 2008, part of the Indian arsenal was de-mated, but a portion is in a 'near-mated condition and almost operational'.[45]

India could take this concept further and move towards a partially mated force with a limited number of weapons fully mated and either in secure storage or deployed on patrol. In an alternative, mated systems could be kept stored and loaded on launchers on a rotating basis. As has been noted above, India is moving towards inducting canister-launched versions of the Agni-V and SLBMs aboard the Arihant class of SSBNs. It is also widely assumed that a canister-launched version of the Agni-IV will eventually enter service. India's arsenal would then consist almost entirely of ready-to-fire missiles. With the submarine force, this will inevitably bring about issues of command and control and launch authorisation which were dealt with earlier.

Canister-launched missiles are effectively always in a ready-to-fire configuration, which would make de-mated warheads completely impractical. This is even more so in the case of SLBMs, which would require their warheads to be mated in advance before deterrent patrols. In both cases, challenges will emerge for refining command and control systems. An alternative deployment practice could be adopted for the land-based Agni missiles whereby a proportion of de-mated missiles and warheads – perhaps even the majority of stock – are kept stored in hardened facilities with only a limited number on patrol. This could allow India to minimise the costs of keeping full operational systems while retaining the survivability and flexibility of its mobile, canister-launched systems. Whether this semi-recessed approach finds favour is as yet unknown, but the option does offer some advantages and circumstantial evidence on the existence of hardened mountain complexes for such systems suggests this is a distinct possibility.

In 2013, Shyam Saran, then a member of India's national security advisory board, made the following remarks:

> If we look at the current status of India's nuclear deterrent and its command and control system, it is clear that at least two legs of the triad referred to in our nuclear doctrine are already in place. These include a modest arsenal, nuclear capable aircraft and missiles both in fixed underground silos as well as those which are mounted on mobile rail and road-based platforms. These land-based missiles include both Agni-II (1,500km) as well as Agni-III (2,500km) missiles. The range and accuracy of further versions, for example Agni V (5,000km) which was tested successfully only recently, will improve with the acquisition of further technological capability and experience. The third leg of the triad which is submarine-based, is admittedly a work in progress. We need at least three Arihant class nuclear submarines so that at least one will always be at sea. Submarine-based missiles systems have been developed and tested in the form of the Sagarika but these are still relatively short in range. It is expected that a modest sea-based deterrence will be in place by 2015 or 2016. There is also a major R&D programme which has been in place since 2005, for the development of a new, longer range and more accurate generation of submarine-based missiles which are likely to be ready for deployment around 2020.[46]

This statement contains the closest 'official' remarks on India's nuclear arsenal for some time. While Saran's comments relating to command and control and survivability are addressed in another chapter, his statements above are noteworthy for openly stating that India is moving inexorably towards a nuclear triad, and if one were to read between the lines, possibly warheads mated on at least a proportion of India's missiles. Also of note is his claim that the Indian arsenal is 'modest'. This is a very vague word – perhaps deliberately chosen – and is also a relative word; 'modest' compared to what? Nonetheless, his statements provide confirmation that India is moving towards a nuclear triad. What this triad will consist of is a matter of extensive debate on both the size of the required arsenal and its composition. It should be noted that much of the discussion on the size of the Indian arsenal has been taking place over a number of years – in some cases a number of decades. The various commentators may not be in tune with the reality of 2021, but their rationales are valid.

The Indian Nuclear Arsenal – How Big is it and How Big Does India Want it to be?

India has had a robust debate on the required size of its arsenal, and this has been supplemented by Western journals. The *Bulletin of the Atomic Scientists* in its 2017 report on India noted that India continued to modernise its nuclear arsenal, with at least four new weapon systems then under development to complement or replace existing nuclear-capable aircraft, land-based delivery systems and sea-based systems. India, in their view, was estimated to have produced enough plutonium for 150–200 nuclear warheads but had likely produced only 120–130. Acknowledging that additional plutonium would be required to produce warheads for missiles under development, the report indicated that India was reportedly building two new plutonium production facilities.[47] Its estimate of India's nuclear forces was as shown in Table 11.

Despite the purported stature of the *Bulletin of the Atomic Scientists* and the credibility given to its findings, there are serious issues with these estimates. As will be discussed later, the Jaguar is most certainly not earmarked for the nuclear delivery role, and while there was mating of the Prithvi SRBM with nuclear warheads during the Kargil conflict, there is no indication of any subsequent employment

Table 11: Indian Nuclear Forces, 2017[48]

Type	NATO Designation	Number of Launchers	Year Deployed	Range (kilometres)[i]	Warhead x yield (kilotons)	Number of Warheads
Aircraft						
Vajra	Mirage 2000H	~16	1985	1,850	1 × bomb	~16
Shamsher	Jaguar IS/IB	~32	1981	1,600	1 × bomb	~32
Subtotal:		~48				~48
Land-based ballistic missiles						
Prithvi-2	n.a.	~24	2003	350[ii]	1 × 12	~24
Agni-1	n.a.	~20	2007[iii]	700+	1 × 40	~20
Agni-2	n.a.	~16	2011[iv]	2,000+	1 × 40	~16
Agni-3	n.a.	~8	2014?	3,200+	1 × 40	~8
Agni-4	n.a.	n.a.	(2018)	3,500+	1 × 40	n.a.
Agni-5	n.a.	n.a.	(2020)	5,200+	1 × 40	n.a.
Subtotal:		~68				~68[v]
Sea-based ballistic missiles						
Dhanush	n.a.	2	2013	400	1 × 12	2
K-15	(Sagarika)	(12)	(2017)	700	1 × 12	(12)
K-4	n.a.	n.a.	?	~3,000	1 × ?	n.a.
Subtotal:		(14)				(14)
Total						~118 (130)[vi]

i Range listed is unrefuelled combat range with drop tanks.
ii US NASIC has estimated the range as 250km (155 miles) but we assume the range has probably been increased to about 350km (217 miles) as stated by the Indian government.
iii Agni I first began induction with the 334th Missile Group in 2004 but did not become operational until 2007.
iv Agni II first began induction with the 335th Missile Group in 2008 but did not become operational until 2011.
v The missile and warhead inventory may be larger than the number of launchers, some of which can be reused to fire additional missiles. This table assumes an average of one warhead for each launcher.
vi The number in parenthesis includes 12 warheads possibly produced for the first SSBN, but not yet operational, for a total stockpile of roughly 130 warheads.

of the missile as a nuclear delivery system. In addition, frequently cited estimates of both the potential yield of India's warheads as well as the numbers of Agni missiles in service lack credibility, with the multiple trials of the Agni-III and Agni-IV not reflected in the number of missiles in service. By the decidedly bizarre logic of the *Bulletin of the Atomic Scientists* team, seven successful Agni-III tests – four of which were user tests – mean only eight missiles are in service and six successful Agni-IV tests – three being user tests – mean that none are in service. These errors render the estimates provided by the team and similar organisations as decidedly unhelpful.

The Stockholm International Peace Research Institute (SIPRI) followed in a similar vein, stating that at the beginning of 2017 India was estimated to have an arsenal of up to 130 nuclear weapons. This represented an increase in the size of the Indian nuclear stockpile from the 110–120 warheads estimated in the SIPRI nuclear data for 2016. India's nuclear doctrine is based on the principle of a minimum credible deterrent and no-first-use of nuclear weapons, but there has been no official statement specifying the required size of the arsenal. India is gradually expanding the size of its

A Prithvi ballistic missile on its Transporter-Erector-Launcher (TEL). (PIB)

A Prithvi ballistic missile during a parade in New Delhi. (PIB)

A test-launch of the Prithvi. (DRDO)

nuclear weapon stockpile as well as its infrastructure for producing nuclear warheads.

While issue may be taken with the numbers, breakdown and details of the SIPRI and *Bulletin of the Atomic Scientists* estimates, given the possibility that India has probably produced fewer than 100 land-based medium and intermediate-range ballistic missiles – none of them as yet with MIRVs – we may reasonably speculate that the number of land-based missile warheads is the same as the number of missiles: i.e. fewer than 100. To these must be added any air-delivered weapons, for which it might be estimated that between 20 and 40 aircraft have been earmarked. This brings us very close to the estimated size of the Indian arsenal given by Western analysts – between 110 and 130 – with the potential that plutonium stockpiles for many more are held and with new production being undertaken each year.

This total is not deemed sufficient to ensure credible effective deterrence – or even credible minimum deterrence. Nevertheless, India is moving inexorably towards a total of some 200 warheads.[49] This would appear to be borne out by both India's existing stockpile and its production capacity of fissile material. This is coupled with the relatively slow – but existing – production of ballistic missiles which, as production of the Agni-IV and Agni-V increases, will add to the number of land-based ballistic missiles available to the SFC. In addition, the pending induction of the INS *Arihant* and its K-15/K-4 missiles will further push the warhead total higher. This has found resonance with analysts such as Brigadier Gurmeet Kanwal, who suggests that with potential losses from a first strike and with a margin of error for reliability, an Indian nuclear arsenal of no fewer than 200 weapons should be considered.[50]

Indian strategic writers, especially those dealing with nuclear issues, have different but usually well-reasoned arguments as to the ideal size and composition of an Indian deterrent arsenal. These range from the most modest figures of the late K. Subramanyam, to a middle path of the modest forces espoused by General K. Sundarji, the perhaps more pragmatic but indubitably larger force levels suggested by such authors as Brigadier Kanwal and Brigadier Nair, and the maximalist and perhaps ideal numbers articulated by Professor Bharat Karnad.

This does not hold true for either larger fission weapons or fusion-boosted-fission weapons which form India's nuclear arsenal. The varying positions taken by analysts on India's nuclear deterrent may thus be divided into the 'minimalist', 'middle path' and 'maximalist' approaches to credible deterrence for India.[51]

The Minimalist Approach

Krishnaswamy Subramanyam was the most articulate advocate for minimalist deterrence. He believed that 'a force of around 60 deliverable warheads could meet adequately India's need for a minimum deterrent'.[52] For delivery of these 60 warheads, Subramanyam advocated the development of 20 Prithvi missiles and 20 Agni missiles. The remaining 20 warheads were slated for delivery by air force aircraft. Subramanyam argued that 'if India were to develop a modest force of 20 Agni missiles, the India–China ratio in deterrence capability will still be higher than the present China–U.S. ratio'.[53] He did not see the need for SSBNs armed with SLBMs and did not make a major a major distinction between low-yield or high-yield weapons, instead emphasising the importance of solid-fuel missiles as delivery systems, stating unequivocally "Whether the warheads are of fifteen kilotons fission or 120–150 kilotons (thermonuclear warheads), both are bound to have a deterrent effect … What is absolutely crucial for credible deterrence is the solid-fuelled missile of appropriate ranges. That is what India needs to concentrate on."[54]

The Subramnayam parameters for nuclear force levels were echoed by Maharajkrishna Rasgotra, a former foreign secretary, who held the view that:

> [S]ome 30 bombs of Hiroshima strength committed against five major targets in Pakistan, 60 deployed against eight to ten targets in China, and another 30 held in reserve for contingencies and deployment at sea, should adequately meet the needs of minimum deterrence. This number (120 warheads in all) allows for possible losses in an enemy first strike and leaves enough for a devastating counterattack.[55]

A Middle Approach

General K. Sundarji, perhaps the most articulate of early Indian nuclear strategists, advocated a pragmatic nuclear force structure of approximately 150 warheads mounted almost entirely on a Prithvi-Agni missile force.[56] Brigadier Vijay K. Nair suggested a force level of 132 nuclear warheads of different types, including weapons in the multi-megaton range – clearly unrealistic for India without a resumption of nuclear tests on a significant basis.[57] Delivery systems would include manned aircraft together with a fleet of five SSBNs with 16 SLBMs on each, and 48 land-based ballistic missiles – 12 SRBMs and 36 medium-range ballistic missiles (MRBMs).

Rear Admiral Raja Menon (retired) suggested that up to 2030, India should maintain an all-missile, land-based force of five regiments with 12 missiles each (survivability being ensured by concealment and rail-garrison mobility), with 50 percent of the missiles having up to four independently targetable warheads per missile. He felt that this arsenal would suffice to withstand the largest possible first strike launched by China and still leave enough missiles remaining to inflict unacceptable damage on China in a second strike. Menon was of the view that India would need a number of hardened silos for additional survivability.

Interestingly, against Pakistan, he proposed a force of 200 cruise missiles, 36 of them nuclear-tipped, because cruise missiles were less provocative – perhaps not adequately considering the relative ease of interception.[58]

The Maximalist Approach

R.R. Subramanian, a senior analyst at IDSA (the Institute for Defence Studies and Analyses) and a physicist by training, held the view that India needed at least 425 warheads if the combined efficiency of the delivery systems is assumed to be 30 percent. By this, Subramanian meant that 30 percent of weapons hit and destroy the targets at which they were fired. So out of 425 warheads, approximately 125–130 could be counted upon to destroy their intended targets. Of these, he estimated that 25–30 would be needed to target Pakistan and about 100 would target China. To assume that India's nuclear force would deliver an overall efficiency of only 30 percent is decidedly pessimistic, though undoubtedly a useful basis for calculations.[59]

Bharat Karnad advocated the need for megaton-class thermonuclear weapons in the Indian arsenal. Professor Karnad suggested that India's primary and secondary target lists could contain about 60 locations in China and Pakistan. In order to ensure, with a small degree of uncertainty, that each of these targets can be destroyed, he recommended targeting each location with four warheads. As it would take time to build a plutonium stockpile, and to design and develop both the Intermediate-Range Ballistic Missiles (IRBMs) and the SSBNs necessary for this targeting plan, Karnad sensibly and realistically suggested that India's nuclear arsenal be gradually built up over a period of three decades to a total of 328 nuclear warheads, as given in Table 12.

Table 12: Requirements for Nuclear Warheads[60]

Timeframe	Maximally Strategic* (warheads)	Minimally Tactical** (warheads)	Total (warheads)
2000–2010	57	30	87
2010–2020	131	40	171
2020–2030	268	60	328

* Intercontinental ballistic missiles (ICBMS), IRBMs, SLBMs, SU-30s armed with N-gravity bombs (NGBs) and N-air-to-surface missiles (N-ASMs) and atomic demolition munitions (ADMs)

** Jaguars and Mirage-2000s armed with NGBs and N-ASMs up to 2010 and SU-30s thereafter

The breakdown of the final figure of 328 nuclear warheads and the proposed delivery systems suggested by Bharat Karnad included:

Four SSBNs with 48 SLBMs (presumably with a single warhead each)
40 SU-30s with 40 NGBs and 40 N-ASMs (maximally strategic) and 30 SU-30s with 30 NGBs and 30 N-ASMs (minimally tactical)
25 ICBMs
40 IRBMs
25 ADMs
50 reserve warheads

Karnad suggested that 253 of the 278 non-reserve warheads should be thermonuclear; the remaining 25 should be atomic demolition munitions. He writes: 'If a counter-cities or counter-value nuclear bombardment strategy is the only one that makes sense, then thermonuclear bombs, with megaton yields, are the most convincing instruments of this strategy.'[61]

This inevitably raises the issue of warhead yields. At present, India's nuclear arsenal perhaps has reliable weapons of the fission-type with yields no larger than 100kt, and at most has fusion-boosted-fission weapons with reliable yields of 200kt, with the possibility of yields of up to 500kt. This renders his otherwise highly feasible force structure as eminently realistic. Of particular note is Karnad's inclusion of no fewer than 140 air-delivered weapons out of the 328 he suggested. This put him at variance with other analysts who have shown a preference for ballistic missiles.

Perhaps the most 'minimalist of the maximalists' was Lieutenant General Pran Pahwa (retired), who recommended in a study for the United Service Institution of India (USI) that India's deterrence should be based on 182 warheads.[62] His very cynical calculations were based on the survivability factor of India's arsenal and worked on the assumption that China was likely to employ two warheads each to destroy every Indian warhead, with efficiency as high as 70–90 percent.[63]

With this calculation, no more than 10–30 percent of Indian warheads would survive a Chinese counter-force first strike. If India had 182 warheads, China would need to fire 364 warheads to eliminate India's arsenal. Given a Chinese arsenal of about 200–250 warheads, a Chinese first strike would leave about 36 Indian missiles unharmed and an equal number of Chinese missiles unlaunched. Since the numbers remaining would be matched, China would be deterred from launching a first strike in the first place. This argument assumes a generous 80 percent success rate for China and does not take into account the possibility that a Chinese first strike is likely to combine counter-value with counter-force targets, thus enabling the survivability of even more Indian warheads.[64]

A close-up view of a BrahMos air-to-surface missile, as suspended under the centreline of the Su-30MKI serial SB 023. Note under the intake, painted in white, an R-77 (AA-12) air-to-air missiles. (IAF)

Another view of a Su-30MKI – here with a 'show-warload' consisting of (from left to right) R-73 training round, R-77 medium-range air-to-air missiles, FAB-250 bombs on multiple ejector racks, BrahMos missile under the centreline, another R-77 and another multiple ejector rack with FAB-250 bombs. (IAF)

Brigadier Gurmeet Kanwal, one of the most nuanced of all India's nuclear thinkers, approached the problem from first principles. He noted that soft area targets, like the population and industrial centres that are likely to be the primary targets of India's countervalue targeting philosophy, would require a much larger number of missiles to destroy with a 90 percent assurance level. To destroy 10 counter-value targets in China, India would need a total of 40 nuclear warheads (at four 200kt warheads per target or 10 20kt weapons per target) to cause unacceptable damage, if the CEP (Circular Error Probable) of the delivery systems was 1,000 metres and an assurance level of about 70 percent was acceptable.

In Kanwal's assessment, India should plan to stock twice the number of land-based warheads and delivery systems that it expects to need. If 25–35 percent of India's deterrence is sea-based, a total of about 150 warheads must be stocked: 120 land-based warheads and about 30 warheads on SLBMs. In addition, India should maintain a prudent level of reserves for larger-than-anticipated damage in a first strike, escalation dominance and unforeseen eventualities. Anticipating escalation dominance and war-termination strategies being dependent on India's ability to launch counter-recovery strikes and fresh strikes if necessary, Kanwal recommended that adding one-third to the previously calculated required number of warheads should be adequate. In total, then, India needed 200 nuclear warheads for a minimum deterrence policy with a no-first-use strategy against China assuming, that this figure would also suffice for deterrence against Pakistan.[65]

In relation to the conventionally accepted figure of perhaps 110–130 weapons being in the current arsenal, it would appear that there is a significant shortfall in respect of any reasonably desirable figure for an Indian arsenal. However, this does not take into account the possibility of either civil reactors being used for weapons-grade plutonium production or the use of reactor-grade plutonium with fusion boosting to eliminate pre-detonation problems. As will be seen later, the option of India diverting one or more of its unsafeguarded nuclear power reactors to weapons material production has the potential to increase India's fissile material substantially.

Tactical Nuclear Warheads?

The consideration given by Indian nuclear theorists and strategists to tactical nuclear weapons – of perhaps sub-kiloton to very low-kiloton yields – being part of the Indian arsenal is very varied. This is partly because of the inherent escalation between tactical and strategic nuclear use being cause for caution, and also because of their questionable military utility to India.

During the Pokhran II nuclear tests in May 1998, Indian scientists conclusively demonstrated the capability to design fractional kiloton weapons. The three smaller, 'experimental' devices had yields of 0.2kt, 0.3kt and 0.5kt, as reported in the May 1998 issue of the *Bhabha Atomic Research Centre* (BARC) *Newsletter*.

Hence, the capabilities declared by Indian scientists must be accepted. It follows that India is in a position to develop tactical nuclear weapons of fractional-kiloton yields, although their manufacture and subsequent maintenance is extremely complex. To this must be added the inherent extreme caution to be exercised in the storage, handling and mating of these tactical nuclear weapons to their delivery systems.

The delivery of tactical nuclear weapons can be done by a combination of manned aircraft, Prithvi and perhaps later BrahMos, Prahaar and eventually Nirbhay missiles. Kanwal has cogently argued that the command and control issues associated with the deployment of tactical nuclear weapons militate against their deployment by India.[66]

Furthermore, their efficacy as retaliatory instruments – given India's no-first-use doctrine – is questionable, as any Indian response to a tactical nuclear strike is to be disproportionate. As such, India has opted against tactical nuclear weapons with good reason.

However, the Indian armed forces have taken steps to prepare themselves for operations in an environment following TNW (Tactical Nuclear Weapon) use by an adversary. This is to be

Newly arrived Rafale fighters. (IAF)

The Rafale is now in service with two IAF squadrons. (IAF)

differentiated from having a proper doctrine for tactical nuclear warfare and is rather part of normal preparedness.

Preparing the Indian Armed Forces for Nuclear Warfare: The Indian Army's Readiness for CBRN warfare[67]

The Indian Army consists of 13 corps-sized formations with a total of 37 divisions and a number of independent brigades. The cutting edge of the Army is centred on three strike corps – each built around an armoured division. The three strike corps – I, II and XXI – are composed of an armoured division (33rd, 1st and 31st, respectively) and two infantry divisions with supporting units. In addition to these formations, the Army has five independent armoured brigades, 15 independent artillery brigades, seven independent infantry brigades, one independent parachute brigade, three independent air defence brigades, two independent air defence groups and four independent engineer brigades.

India has understood that its previous operational doctrine, which emphasised the use of its three strike corps, was inadequate to meet the changing realities of the India–Pakistan battlefield. To this end, the previously designated 'Holding Corps' along the India–Pakistan border have been renamed as 'Pivot Corps'. They have been reinforced with the addition of eight Integrated Battle Groups (IBGs), which are of division size and combine mechanised infantry, armour, artillery and aviation assets. The IBGs are to work in conjunction with the Indian Air Force to conduct offensive military operations and be available for operations within 48 hours of the order being given. This reorientation of assets has increased the offensive capability of previously defensively orientated formations and has also made them considerably more flexible.

To enable India to call Pakistan's nuclear bluff, and further to enable its combat formations to operate effectively under the backdrop of TNW use, the Indian Army has to give priority to passive and active CBRN measures. Passive CBRN measures involve protecting personnel and equipment from the effects of CBRN use, while active measures involve the interception of the delivery systems of CBRN agents before they can inflict damage on their intended targets. The extent to which India has undertaken such measures is discussed below.

Passive CBRN Defensive Measures

India's 63 armoured regiments and 26 mechanised infantry battalions have significant levels of protection from CBRN attack. The Arjun, T-90 and T-72 tanks all have CBRN protection systems, while the BMP-1 and BMP-2 force that conveys the mechanised infantry units also have such systems. Furthermore, each BMP comes with two TDP-1 decontamination kits. It is, however, unclear if India has maintained and replaced life-expired filters on these vehicles or refurbished and/or replaced ageing decontamination kits. In addition, India has also deployed a number of indigenously developed CBRN reconnaissance and decontamination vehicles. Of note, however, is its continuing reliance on towed artillery, which is far more difficult to operate in CBRN conditions. The lack of appropriate self-propelled artillery is also a major deficiency. The latter point cannot be overemphasised, as trying to operate towed artillery in a CBRN environment is not only difficult but could place an excessive physiological strain on artillery crews operating in high ambient temperatures. That said, respiratory protection is feasible, as even the primitive and claustrophobic respirators used in the First World War were employed by artillery crews during chemical bombardment in hot summers, which although uncomfortable, were not debilitating.

Training of Indian Army troops in full NBC suits. (Indian Army)

Indian troops on exercises wearing locally made NBC suits and locally made respirators based on the Kareta Nova design. (PIB)

With respect to infantry formations, in the 1980s the Indian Army began some tentative preparations for CBRN warfare. A quantity of CBRN equipment was imported from the USSR, but these proved to be useless in Indian environmental conditions. Following that, in 1987, the Defence Research and Development Organisation produced prototypes of CBRN suits, decontamination suits, facelets, overboots and CBRN tents. This equipment has entered production and is now in service with the armed forces. From 1987 onwards, the Army, through its College of Military Engineering, began running familiarisation courses in CBRN warfare, while DRDO scientists have conducted theoretical courses at the brigade level. Over time, this became practical training.

The Ministry of Defence has allocated priority to the indigenous production of CBRN gear, which has also involved the participation of the private sector in the manufacture of CBRN suits. India seems to have purchased 50,000 S6 and S10 respirators and No.1 Mk3 suits from the UK in the 1980s (the types being identified through photographic evidence), 150,000 Draeger Kareta Nova respirators in the early 1990s, and subsequently a larger order of 334,000 full sets of Individual Protective Equipment (IPE) with CBRN suits and locally produced variants of the Kareta Nova respirator.[68] Further purchases have included 40,000 new-model CBRN suits in 2011.[69] CBRN IPE has a limited shelf life; in order to retain CBRN readiness levels, India would need to replace life-expired items, refurbish them where possible and ensure proper storage of equipment. The latter point is of particular importance given India's climate, which could be detrimental to the integrity of CBRN respirators if special care is not taken in storage, preventative maintenance and periodic thorough refurbishment. A recent order for more than 30,000 sets of IPE from the United States is perhaps the precursor to the adoption of a new model of IPE gear.[70]

To the extent that research has thus far permitted, it is suggested that, at present, India has an adequate stock of CBRN gear for individual protection of its infantry units. Older equipment is now making way into training establishments and new production replacing IPE sets that have become life-expired. Decontamination equipment, locally designed and manufactured, is also available in some quantity, but on a scale that is far from lavish. However, it is a matter of conjecture whether the Army's towed artillery crews are issued with or trained to operate in CBRN gear.

Despite all the above, the exact status of CBRN preparedness and training in the Indian Army remains unclear. For instance, it is not known whether training and equipment has gone down to the battalion level. It is certainly the case that outside of CBRN exercises, Indian troops do not carry CBRN haversacks. Since the late 1990s, some exercises have been conducted with troops in full CBRN gear and decontamination vehicles in operation – the latest being in April 2016. However, these exercises have not been widely publicised and their extent is not clear. Nevertheless, it is clear that the foundations for reasonably strong passive CBRN measures have been laid in the Indian Army. It remains to be seen if the threat of Pakistani TNWs will provide further impetus in this regard. It should be noted that the Indian Air Force has also initiated measures to operate in a CBRN environment. Some decontamination exercises have been carried out, and recently a repair and refurbishment facility has been established for CBRN gear. However, once again, it is unclear if these efforts amount to anything more than basic familiarisation measures or a much more serious impetus towards protecting aircrew as well as ensuring effective IAF operations in a CBRN environment.

Delivery Systems – The Triad Emerges

India has moved towards the development of a triad based on a combination of delivery systems which include manned aircraft, land-based ballistic missiles and, more recently, submarine-launched missiles.

Manned Aircraft

The Indian Air Force, with a force of some 260 Su-30MKIs, 50 Mirage 2000s and over 110 Jaguars, has many options for manned aircraft delivering nuclear gravity bombs. The weight, yield and dimensions of Indian gravity and/or glide bombs are not known, but it might be anticipated that Indian weapons range between 250kg and 1,000kg and can have yield of between 15kt and maybe as much as 200kt – the latter being fusion-boosted-fission weapons.

Indications are that the first Indian nuclear weapons design had a mass of about 1,000kg, with a yield of 12–15kt. However, subsequently, perhaps by 1982 (when rumours of a fresh round of nuclear tests were being circulated), these weapons had been scaled down to a more manageable mass of between 170kg and 200kg.[71] Later, it appears that a 100kt fission weapon was produced for aerial delivery with a mass of 200–300kg.[72] If this is accurate, it would mean that India had perfected a relatively high-yield fission weapon with a relatively low mass for its class. It should also be noted that by the 1990s, fusion-boosted-fission weapons had become available for air delivery.

While India had developed a stockpile of some 15–20 nuclear gravity bombs in the period from 1974–1985 – all of which were in the 15–20kt range – the country became an effective nuclear weapons state in May 1994. It might have been thought that this milestone would have been achieved earlier, and indeed it was certainly possible for this to be done. However, mating of the nuclear weapons with the delivery aircraft proved to be challenging.

India's first deployed air-delivered nuclear weapons were earmarked for use from the Jaguar strike aircraft. Initial tests with the Jaguar IS – India's principal deep-penetration strike aircraft – found that the ARDE (Armament Research and Development Establishment) bomb-pods were too heavy for the Jaguar; once fitted to the centreline pylon of the aircraft, the ground clearance of a mere 2 inches was deemed unsafe.[73]

India then chose to use its Mirage 2000H/TH fleet for the nuclear delivery role.[74] At least one of India's three IAF Mirage 2000 squadrons was tasked with the nuclear strike role – although it does not seem that this was its exclusive role, given the limited number of aircraft available. One might expect that the 36 Dassault Rafales may find themselves supplementing and thereafter supplanting the Mirages in the nuclear strike role.[75] It has also been speculated that the Su-30MKI fleet may have a nuclear strike role. This would not be unrealistic, as the greater payload capacity offered by the Su-30MKI could eventually permit the carriage of air-launched nuclear-armed cruise missiles in the future. It has even been suggested that the

A Su-30MKI in the process of landing after another test flight with the BrahMos missile. (DRDO)

Initially at least, fighter-bombers were the primary means of delivery of Indian nuclear weapons. The first platform associated with such armament was the Mirage 2000. India originally acquired 36 single-seat Mirage 2000H interceptors and four two-seat Mirage 2000TH conversion trainers. They were operated by Nos 1 and 7 Squadrons as interceptors, primarily to counter the US sale of General Dynamics F-16s to Pakistan in 1980. The versatile weapons system of the Vajra, as the type is nicknamed in IAF service, has enabled multiple upgrades since the purchase, converting the type into one of the IAF's most potent strike platforms. Nine additional Mirage 2000s were acquired between 1986 and 1994, and a third order followed in 2004, bringing the total to about 50. Since 2011, the entire fleet has been undergoing an upgrade to the Mirage 2000-5 Mk. 2 standard (locally designated Mirage 2000I/TI), making them compatible with a wide range of advanced weaponry. (Artwork by Tom Cooper)

Presently, the most important fighter-bomber in the Indian arsenal equipped to deploy nuclear weapons is the Su-30MKI. Since 1997, 272 such jets have so far been acquired, including 50 from Russian production and 222 licence-built by HAL of India, while an additional 12 may be on order. As of 2021, Su-30MKI Sqns are 2,8, 15, 20, 24, 30, 31, 102, 106, 220, 221, 222 and TACDE (Tactics and Air Combat Development Establishment). Up to 40 Su-30MKIs have been modified for the deployment of nuclear weapons. This illustration depicts the example registered as SB 040 (from the fourth batch, delivered in 2004), armed with the BrahMos hypersonic cruise missile – carried under the centreline, between the engine nacelles – which has been undergoing testing since 2017. Air-to-air missiles such as the R-73E (wing-tips), R-27ET (underwing pylon) and R-77 (or the indigenous Astra missile) would be carried for self-defence purposes during any nuclear mission. (Artwork by Tom Cooper)

Following an order from 2016, the Indian Air Force received the first of 36 Dassault Rafale multi-role fighters in 2019. Deliveries of a total of eight two-seaters (Rafale DH) and 28 single-seaters (Rafale EH) were due to be completed by January 2022. The Rafale DH, the two-seat variant illustrated here, is likely to serve as a conversion trainer, but it also has an important strike role – even more so because it can be armed with the MBDA SCALP-EG subsonic cruise missile (shown on the outboard underwing pylon), capable of carrying a nuclear warhead. For self-defence purposes, the type can be armed with Meteor, MicaIR and Mica EM air-to-air missiles: it is expected that in the future, Rafales and Su-30s will also receive the indigenous Astra air-to-air missile. (Artwork by Goran Sudar)

Perhaps unusually, the Indian Navy deploys two Sukanya-class off-shore patrol vessels – INS *Subhadra*, shown here, and INS *Suvarna* – equipped with storage, elevation and launching facilities for the Dhanush missile, which has a range of 350km. The Dhanush is the ship-launched variant of the Prithvi ballistic missile, possibly having a slightly longer range at the cost of a reduced payload. Their very light self-defence armament indicates that both vessels are test-rigs used to develop elevation, stabilisation and launch mechanisms for sea operations, necessary for K-4 and K-15 missiles. (Artwork by Ivan Zajac)

In 2009, India launched the INS *Arihant* (S-2) – the lead vessel in a class of three or four SSBNs. The second submarine of this class, INS *Arighat* (S-3), followed in 2017, and is yet to be officially commissioned into service. The Arihant-class not only completes the nuclear triad of the country, but also provides survivable platforms for a possible second strike. Displacing around 6,000 tonnes, they are relatively small SSBNs, but still carry powerful armament, including either 12 K-15 or four K-4 ballistic missiles with a maximum range of around 3,500km. A follow-up class of slightly bigger SSBNs is expected to carry between eight and 12 K-4s, while the development of longer-ranged weapons – like the S-5 and K-6 – is reportedly underway. (Artwork by Ivan Zajac)

By 1988, relatively few of the 14 Breguet Br.1050 Alizé anti-submarine warfare aircraft acquired and operated by INAS 310 were still fully mission capable. While originally serving aboard INS *Vikrant* (the first aircraft carrier of the Indian Navy), from 1987 they were relegated to shore-based operations and were usually deployed in support of the Indian Peace Keeping Force on Sri Lanka. In that area of operations, they proved highly successful in combating speedboats carrying supplies for the LTTE, initially with up to six (three per wing) 127mm unguided rockets, later with Matra F4 pods for 14 68mm unguided rockets. Camouflage colours of Indian Alizés included extra-dark sea grey on top surfaces and Sky, Type-S, on undersurfaces and sides. National insignia was worn in six positions, with service title (in black) on the rear fuselage, followed by the full serial underneath the horizontal fin (the last two figures of the latter were repeated below the cockpit). A small unit insignia was often applied on the rudder. (Artwork by Tom Cooper)

Starting in 1969, India acquired a total of 15 Westland Sea King Mk.42/42A and 20 Sea King Mk.42B anti-submarine helicopters, along with six Sea King Mk.42C SAR and utility helicopters. They were operated by INAS 330 and INAS 336. As of 1988, all were painted as shown here, in white on top surfaces and dark blue everywhere else. National markings and serials were applied on the rear of the fuselage, service titles on the boom and a white 'W' was frequently applied further forward, indicating their frequent deployment aboard INS *Vikrant*. The Sea King Mk.42B variant could be armed with Sea Eagle anti-ship missiles, but during Operation Cactus, no such armament was necessary: instead, Indian Sea Kings were deployed to search for and track ships used by the LTTE. (Artwork by Goran Sudar)

Hindustan Aeronautics Limited (HAL) manufactured over 300 Aérospatiale SE.316B/C Alouette III helicopters under licence, under the designation Chetak. The majority of these were acquired by India's armed forces, including the Indian Navy, which equipped the INAS with them. Together with Sea Kings, a handful of these – all operated by INAS 321 – played a crucial role in tracking down and assaulting ships used by the LTTE for its attack on the Maldives. As far as is known, all were painted in dark blue overall, wore national markings and white serials on the rear of the hull, and service titles in white on the boom. Most of INAS 321's Chetaks were equipped with rescue winches but could also be armed with machine guns or torpedoes, and frequently had the left cabin doors removed for easier access. (Artwork by Tom Cooper)

In the early 1960s, the Indian Army acquired 3,500 Nissan Patrol 4WD utility vehicles. These proved so highly popular in service that licenced production of the Nissan Patrol P60 was launched at the Vehicle Factory Jabalpur (VFJ). The acronym for the military division of the VFJ – Jabalpur Ordnance and Gun-carriage Assembly – eventually resulted in the name of the Indian-manufactured version: Jonga. This robust 4WD, powered by a 4.0-litre, in-line six-cylinder 145hp engine, was deployed as a six-man personnel carrier, an ambulance and a reconnaissance and patrol vehicle: the latter two are illustrated here, one with the FN MAG general purpose machine gun installed in front of the co-driver and the other with the same machine gun on a pedestal in the rear. (Artwork by David Bocquelet)

The final role in which the Jonga was deployed by the Indian Army was that of a gun carrier: as such, the vehicle was armed with a single M40 106mm recoilless gun, installed as shown here. The 2nd Para deployed a battery of four gun-carrying Jongas during Operation Cactus. As far as is known, these received a relatively simple camouflage pattern, consisting of wide stripes of sand applied on top of the olive green overall livery. (Artwork by David Bocquelet)

Based on a Canadian design, the 75/24 Pack or Mountain Howitzer was developed in a cooperation between the Armament Research & Development Establishment (ARDE) and Ordnance Factories to become the first indigenously designed and manufactured lightweight howitzer in Indian service. Its nomenclature is based on the calibre; 75mm, with a barrel length of 24 calibre. Although taking a long time to develop, and proving something of an 'underdevelopment', it became highly popular once it entered service, and at least one such battery was deployed during Operation Cactus. (Artwork by David Bocquelet)

This illustration shows Brigadier F. F. C. Bulsara at Hulhule International during the Indian intervention in the Maldives. He wears a uniform with non-regular pattern, most likely privately purchased. Note the only visible insignia on his fatigues: the parachute wings above the right pocket, identifying him as a qualified paratrooper. His maroon beret has the gold insignia of a senior officer of the Indian Air Force. (Artwork by Anderson Subtil)

This private of 6 Para Battalion of the Indian Army is shown wearing a maroon beret with the cap badge of the Parachute Regiment and the Indian version of the DPM camouflage uniform. He is armed with the Indian-made version of the SLR 7.62mm assault rifle (magazines for which were carried in a single canvas pouch attached to the Model 1937 belt) and is shown carrying a box of 2-inch mortar bombs on his shoulder. (Artwork by Anderson Subtil)

In the case of the Tamil guerrilla of 1988, there little to say about 'uniforms': they simply wore common clothes of that time. However, this individual is shown wearing a tactical vest (probably stolen from Indian Army stocks), with various pouches for ammunition magazines and grenades. Notable is the small Motorola radio (just visible behind the breach of the AK assault rifle), used for short-range communication. (Artwork by Anderson Subtil)

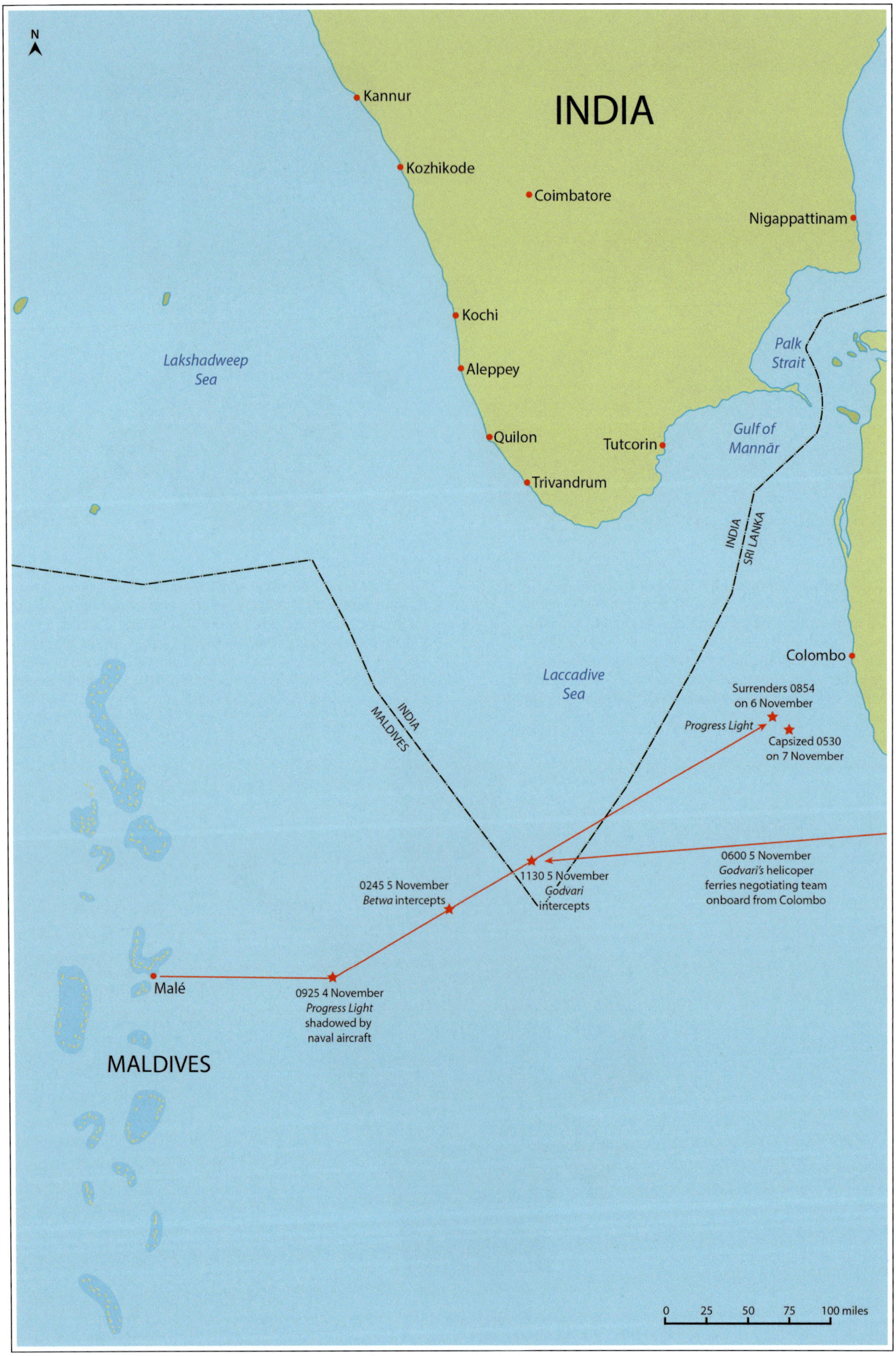

A map of Operation Cactus, in November 1988, depicting the most important operations carried out between India, the Maldives and Sri Lanka. (Map by George Anderson)

The major types of ballistic missile in Indian service are, from left to right (and shown in a variety of colour schemes), the Prithvi, Agni-I, Agni-II, Agni-III, Agni-IV (formerly Agni-IIAT), Agni-V, and K-15. The first six are land-launched, and the last – also named the B-05 or Sagarika ('Oceanic') – is submarine launched. All can carry nuclear warheads of about 1,000kg mass. (Artworks by Tom Cooper)

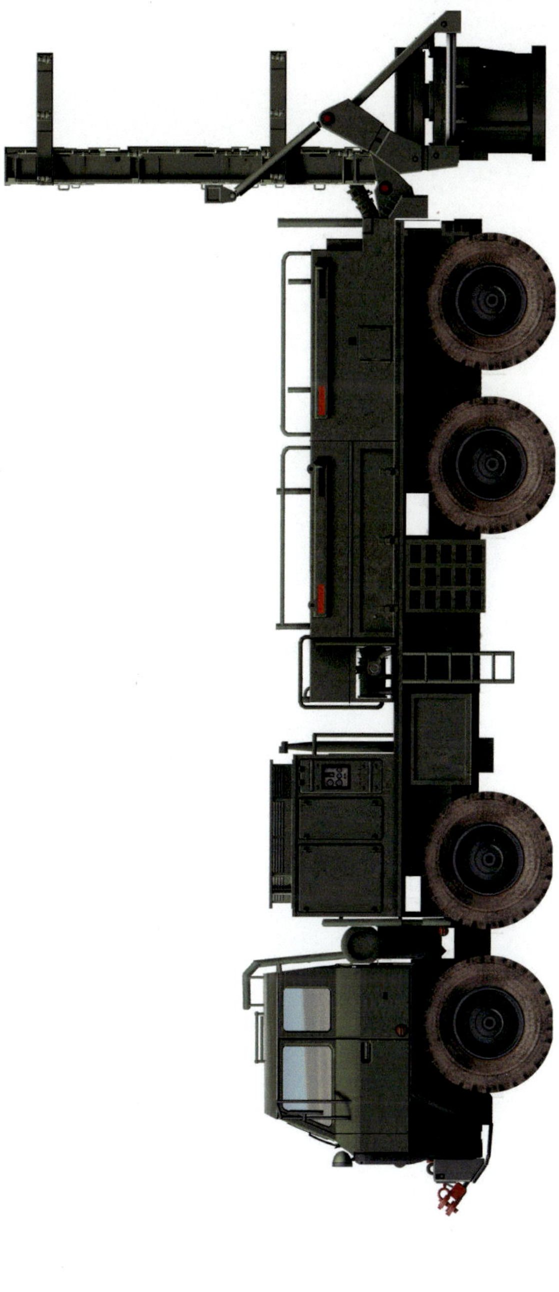

India has developed custom-made Transporter-Erector-Launchers (TELs) for all the ballistic and cruise missiles in its service, enabling these to be quickly transported and fired from almost anywhere. This illustration shows a TEL based on the chassis of a commercial truck made by Tata Motors Ltd of Mumbai, used for deployment of the Prithvi I/II/III family of missiles, the Prithvi Air Defence missile and – possibly – for test-launches of the PDV. (Artwork by David Bocquelet)

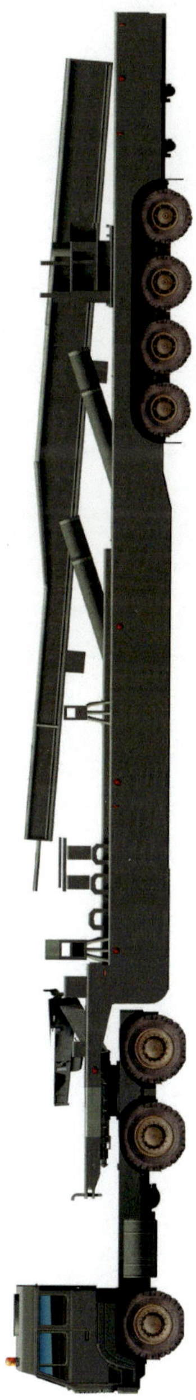

A reconstruction of the truck-towed TEL for the Agni-III ballistic missiles. The Agni-III is 17 metres long with a diameter of 2 meters, weighing 48 tonnes, and can reach targets up to 5,000km away from the launching point. The latest Agni-IV and Agni-V are strategic nuclear weapons with a range of up to 12,000km. Due to the sheer size and weight of the weapon, TELs of this kind are limited to deployment along India's road network, while the Agni-II is launched from a specially modified railway car. (Artwork by David Bocquelet)

(Map by George Anderson)

A frontal view of one of the Rafale B fighters in the process of entering service with the Indian Air Force. (IAF)

Su-30MKI has now supplanted the Mirage 2000 in the nuclear strike role.

It is believed that certain squadrons have been given such a tasking in wartime, with at least one squadron of each of the Su-30MKIs and Mirage 2000s being so employed. It might be expected that a degree of additional Electro Magnetic Pulse protection would be provided to these aircraft. However, given the fact that it is highly unlikely that whole squadrons can be permanently assigned to the nuclear strike role without compromising India's ability to conduct air operations in the event of hostilities, determining which aircraft need to be given additional EMP shielding presents a challenge. It might be necessary for India to consider additional EMP shielding for a larger proportion of its air assets than would be necessitated by the number of nuclear weapons earmarked for aerial delivery.

As to survivability, India has constructed underground hangars in at least three airbases – Adampur, Hindon and Bareilly.[76] However, surface shelters in the form of hardened aircraft shelters are not yet available for most of the Su-30MKI fleet. The building of 108 new generation hardened aircraft shelters with improved protection characteristics is currently in progress at the time of writing, with the first batch of 36 to be completed within the next five years.[77]

Cruise Missiles

Besides its ballistic missiles, India has a cruise missile programme which, when completed, would give the country another option for the delivery of nuclear warheads. It must be clarified that India's cruise missile projects have to be divided into two – the BrahMos supersonic system and the Nirbhay subsonic system. While there have been multiple reports of the BrahMos being nuclear capable and/or fitted with a 20kt warhead, it is suggested that this may not be entirely accurate – although the potential certainly exists for such an event.[78] Indeed, the BrahMos is set to experience something of a transformation. Initially claimed to be a missile with a 290km range, India's accession to the Missile Technology Control Regime enabled the system to be tested to its full range of 450km against both land and sea targets. It is anticipated that a longer-ranged BrahMos – perhaps capable of ranges up to 900km – will be developed in air-, sea- and land-launched versions. It is possible that a nuclear warhead could be developed for such an extended range version, although the principal task of the BrahMos would be long-range precision attack.[79]

The Nirbhay, however, is intended to be nuclear capable, although it will also be an important component for convention payload delivery. The Nirbhay is India's first strategic cruise missile, with a range exceeding 1,000km, a speed of between 0.8 and 0.9 Mach (speed of sound), a launch weight of 1,500kg and a length of 6m. The payload size of the missile is not yet known, but there is speculation that it would be capable of carrying either conventional or nuclear warheads of 24 different types. It is intended to fly at very low altitudes of 20m or less to evade radar detection, using a terrain-following function which enables it to use mapped contours of the ground for both concealment and updating its inertial navigation system.

In contrast to ballistic missiles, cruise missiles can be cheaper, are much slower and use a different flight path entirely. They are also more flexible: in the South Asian context, ballistic missiles are inextricably linked to nuclear weapons and as such cruise missiles attract much less attention, making them useable weapons for the delivery of conventional warheads while retaining the ability to carry a nuclear payload.

Compared to the BrahMos missile already in service, the Nirbhay is longer-ranged (1,000km+ vs 290km) but slower, and the BrahMos is considerably longer (8.4m) and heavier (3,000kg). However, while the BrahMos is intended to cruise at higher altitudes and then complete its terminal run to target at a very low altitude of 10–15m, the Nirbhay can complete much of its flight at these lower altitudes. The Nirbhay also has the ability to loiter and to pick out and engage a specific target in a multiple-target environment.[80] In addition, it is very unlikely that the BrahMos has, at this stage, any ability to carry a nuclear warhead – its manufacturers specify a conventional payload only – although there has been some informed comment to the contrary. The Nirbhay has been tested four times to date, with

A test-drop of a BrahMos air-to-surface missile from a Su-30MKI fighter-bomber during certification testing. (DRDO)

Test-launch of a Nirbhay missile on 15 April 2019. (DRDO)

Another look at the ascending Nirbhay missile, during a test launch on 7 October 2014. (DRDO)

three tests being failures and the other termed either a partial or complete success depending on which report is read.

On 12 March 2013, during its first test, the missile took off from the launch pad successfully and reached the second stage of propulsion, travelling 15 minutes along its predicted path, covering 30 percent of its intended 1,000km at a speed of 0.7 Mach. It then veered away from its trajectory, forcing the missile to be destroyed

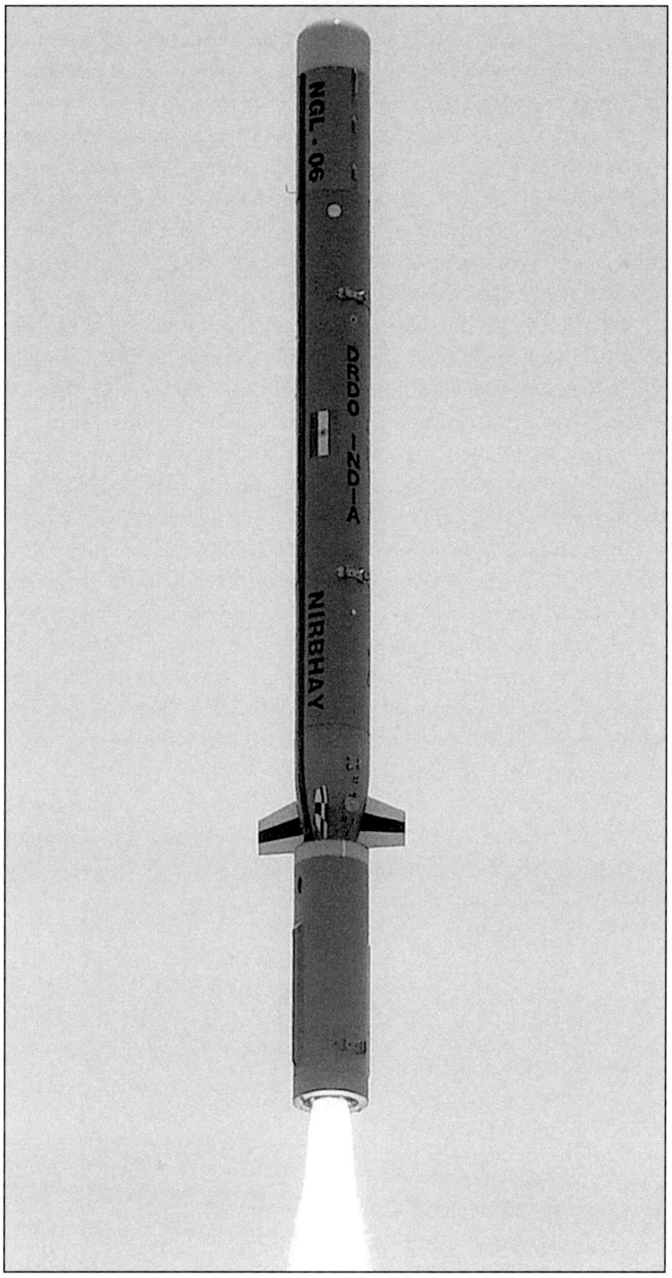

Details of the Nirbhay missile seen on 15 April 2019. Note the configuration generally resembling that of the US-made BGM-109 Tomahawk Land-Attack Missile (minus wings) and the booster section. (DRDO)

in mid-flight. On this occasion, it was suggested that the fault was a defect in the inertial navigation system.[81]

On 17 October 2014, the second test of the missile was much more successful. All the test parameters were purportedly met and the missile completed all 15 way-points. The missile covered a distance of more than 1,000 km in a flight that lasted for over an hour and 10 minutes. It should be noted, though, that the missile did not manoeuvre as desired at low altitude. While the missile test was, therefore, largely successful, it was not an unqualified success and should have raised questions about the missile's low-altitude performance. Nevertheless, the test demonstrated the core capability of the Nirbhay missile.[82]

The third test of the missile took place on 16 October 2015. The missile was to be tested for its low flying capability, being brought down from 4,800,m to 20m gradually and in stages. All initial critical operations such as booster ignition, booster separation, wing deployment and engine start were successfully executed, and the Nirbhay reached the desired cruise altitude. However, 11 minutes into the flight, after covering only 128km, the missile crashed into the Bay of Bengal.[83]

The fourth test of the missile took place on 21 December 2016, when there was once again a successful booster ignition and launch. Less than four minutes after lift-off, however, the missile veered off its intended trajectory and outside the safety zone, leading to it being destroyed in flight. Initial reports suggest the flight control hardware and software failed to actuate the missile control services and that the engine lost thrust two minutes after take-off.[84]

Apparently stung by criticism that the third and fourth tests were rushed, the fifth test of the missile took place on 7 November 2017. This time the test was successful, and was followed nearly two years later by a sixth test, a successful trial which took place on 15 April 2019. This time the test missile covered some 650km, and navigated way-points located at altitudes varying from 5m to 2.5km.[85] This trial completed the developmental trials of the missile, thus paving the way for user trials by the military. In a vote of confidence in the system, the Indian Army deployed a limited number of conventionally armed Nirbhay missiles along the Line of Actual Control (LAC) separating Indian- and Chinese-controlled land in the Ladakh Union Territory.[86]

Whether either the Nirbhay or the BrahMos has a nuclear role is as yet unclear. However, given India's progress in miniaturising warheads, it is possible that such a warhead can be adapted for use by these missiles. However, of equal note is the fact that it is probable that both the Nirbhay and the BrahMos will have primary conventional roles.

While an air-launched version of the BrahMos is now in service, complementing both land-based and ship-launched variants, the Nirbhay's fate is less clear. All tests have been from a mobile, land-based launcher, but both the IAF and Indian Navy have indicated a degree of interest in procuring the system for their use.

In the case of the IAF, it is expected that an air-launched version would be carried by the Su-30MKI on a centreline pylon. If the Nirbhay retains its 1,000km range, it would make it an excellent weapon for possible employment as a nuclear-armed cruise missile, thus totally changing the dynamic of air-delivered weaponry in respect of the Indian nuclear arsenal. The Indian Navy is understood to have requested an increase in the range of the missile to 1,500km for launch from surface ships, with as yet no indication regarding deployment from submarines, either nuclear or conventional.

India's Ballistic Missiles – The Backbone of the Arsenal[87]

India's first operational ballistic missiles of the Prithvi SS-150, SS-250 and SS-350 variants were relatively widely produced, with perhaps 180 of the SS-150s and 60 of the SS-250s and SS-350s.[88] However, it is less clear whether any of these variants has had anything more than a secondary nuclear role, with anecdotal evidence suggesting that four SS-150s were armed with 15kt fission warheads during the 1999 Kargil War.[89]

Development of the missile began in 1983. In its current configuration, the missile is 8.56m long, 1.1m in diameter and weighs 4,000kg. Unlike later Indian missiles, the Prithvi family uses a single-stage, liquid propellant engine, which is essentially two liquid propellant motors side-by-side that provide aerodynamic control as well as thrust vectoring. This engine control allows the missile to stop climbing when it reaches an altitude of 30km, travel horizontally at this altitude and dive on its target at an 80° angle.

The missile has a minimum range of 40km and, in its later variants, a maximum of 350km.

The missile has a reported accuracy of 50m CEP (Circular Error Probable) against targets at 150km. At present, it uses an inertial guidance system and is launched from a Transporter-Erector-Launcher (TEL) vehicle. It takes approximately two hours to prepare a missile for launch, due to pre-flight fuelling and various support functions that must be accomplished. Two TEL vehicles require an average of 18 support vehicles that include warhead carriers, power suppliers, fuel carriers, cranes and others.

The Prithvi-I's first test flight was in 1988 and it officially entered service in 1994. By 1999, there had been 16 successful tests of this missile system. Further trials at user level were conducted from 2003. The missile was first deployed with the 333rd Indian Army Missile Group. Thereafter, the Indian Army raised two more groups: the 444th Indian Army Missile Group and the 555th Indian Army Missile Group. Though the Indian Army intended to deploy the Prithvi as a viable tactical weapon with conventional warheads, more recent tests of the system have been reportedly conducted by the Strategic Forces Command. It should also be noted that the Indian Air Force operates a small number of Prithvi SS-250 variants, while the Navy, with its Dhanush missile, has two vessels equipped to launch these missiles.

Test-launch of an Agni-II missile during user trials. (DRDO)

However, the Indian land-based nuclear deterrent rests upon the five variants of the Agni family of ballistic missiles, which range from medium-range ballistic missiles (Agni-I) to limited-range intercontinental ballistic missiles (Agni-V). All bar perhaps the Agni-V are in operational service, with numbers as yet unknown but with production ongoing, albeit at an apparently low rate.

India's testing of missiles follows guidelines established in 2002 after statistical analysis by a former chairman of ISRO (the Indian Space Research Organisation), K. Kasturirangnan, who conducted statistical analysis of failures of major systems, components and subsystems, plus their performance in static tests. His conclusion was that two or three consecutive successful test-firings would meet the standards of reliability within the constraints of affordability.[90]

The test of the Agni-V missile in its final deliverable configuration on 26 December 2016 represented a major step forward for India's strategic deterrence posture. Lost in the very legitimate emphasis on what the test means for India's nuclear deterrent, is the fact that the Agni-V represents the culmination to date of an evolutionary process in the development of India's ballistic missile capability, whereby one phase of development drew heavily on those versions of the missile that came before.

There is one important point to note: India has used depressed and lofted trajectories to simulate ranges on it missiles. Details of the trajectories have not been disclosed, extrapolations having

An Agni-I missile soaring into the skies during a test launch on 13 July 2012. (DRDO)

been made from the NOTAM warnings issued to pilots alerting them to potential hazards along a flight route. This leads to a strong suspicion, with some foundation, that India has understated the range of its missiles from the Agni-II onwards. This trend continues, with the words 'more than' being almost a permanent prefix to range figures of Indian missiles. Even more perplexing than the range estimates are the number of missiles produced and in service – particularly given the number of user trials from production batches.

An Agni-II missile on its dedicated TEL during the Republic Day Parade in New Delhi in 2004. (PIB)

The Agni Series of Missiles

The original Agni missile – formerly a 're-entry demonstrator' and which we can now term the Agni-TD – was tested on 22 May 1989. By its third test in 1994, the 2,500km-range missile had achieved a demonstrated range of 1,450km, with a claimed CEP of 300m.[91] However, the most remarkable aspect of this progenitor to the Agni missiles of the twenty-first century was that it was a hybrid, using the liquid-fuelled Prithvi-1 missile as the basis for its first stage, coupled to a solid fuel booster reminiscent of, but not identical to, the booster of ISRO's SLV-3 launch vehicle. The Prithvi, which by 1994 had entered bulk production, was a reliable system, with propulsion based on India's stillborn 'Devil' SAM project of the 1970s and married to the proven SLV-3 booster. This low-risk approach enabled India to develop an Intermediate-Range Ballistic Missile (IRBM) capability at a relatively low cost, using proven technology.

However, it was soon realised that the Agni-TD was not a practical weapon, owing in large part to its hybrid propulsion and the inherent problems of keeping such a system in a state of readiness, or at least in a state of near-readiness. Moreover, it would appear that India, learning the lessons of the 1980s, decided against deploying its land-based deterrent in static hardened silos and opted for mobile systems. The Agni-TD was not a mobile system, and developing its successor required not only an improved missile, but the development of a mobile launcher that could be deployed operationally. It should be noted that while unlikely, the Agni-TD could have been weaponised. Admittedly, it was far from an ideal system, but it appeared that its designers were always cognisant of the possibility of the need to weaponise it.

An early example of the Agni-II missile in erected position, ready for launch. (DRDO)

The Agni-II

First tested on 11 April 1999, the Agni-II had a range of anywhere between 2,000km and 3,000km, with all solid-fuel propulsion. Tested to a range of 2,300km with a 1,000kg payload in 1999, the Agni-II established a pattern for Indian missiles – production and user trials would follow three consecutive successful technical trials.[92] The 1999 test was followed by one on 17 January 2001 and a third on 29 August 2004. Production of the Agni-II followed thereafter, but it was not until 2009 that user trials began, the first two of which were complete failures due to quality control problems during manufacture. However, on 17 May 2010, the first user trial from a production batch was successfully completed, effectively marking the missile achieving operational status within the Indian arsenal.[93] This was followed by similar production batch trials undertaken by the SFC on 30 September 2011,[94] 9 August 2012, 7

April 2013[95] and 9 November 2014.[96] This resulted in the Agni-II becoming the backbone of India's IRBM force. However, a test on 4 May 2017 failed. A subsequent test on 20 February 2018 by the SFC was successful.[97] This was followed by a test on 16 November 2019, again by the SFC, which once again was successful.[98]

The Agni-II was India's first viable production IRBM. It is mounted on a rail-mobile TEL, with some suggestions that the TEL can also be made road-mobile. In either case, the TEL provides the Agni-II with flexibility in deployment as well as survivability. Furthermore, the system can be deployed within 15 minutes. The Agni-II also marked India's departure from liquid propellants to all solid-fuel systems. Furthermore, the Agni-II's reputed CEP of between 30m and 100m at maximum range makes it an accurate system. Production of the Agni-II has been reported since 2001 and 2002, but it is unclear as to what level of production has been achieved, with estimates ranging widely. It is to be noted that in 1998, instructions were purportedly given to prepare production facilities for the Agni-II at the rate of one per month,[99] but it is not certain whether such production levels have ever been achieved.

The Agni-I

In a curious twist, the Agni-II, which evolved from the Agni-TD, spawned the Agni-I. This 700km-range MRBM was borne out of an operational necessity for a missile with a longer range than the Prithvi to serve as a deterrent against Pakistan. The Agni-II was effectively re-engineered to create a smaller, lighter and road-mobile system that is both survivable and easily deployable. The Agni-I, therefore, could (at least conceptually) be deemed a single-stage Agni-II. Since its first test in 2002, it has undergone a series of user trials from 2007 onwards, the last being on 22 November 2016.[100]. While the missile itself is an achievement of note, the road-mobile launch system should be considered of equal importance as it has evolved into the road-mobile TELs for the Agni-IV and Agni-V.

The Agni-P

On 28 June 2021, India successfully tested the successor missile to both the Agni-I and Agni-II systems. The new missile was designated the Agni-P (Agni Prime). The canisterised missile has a range of up to 2,000km and looks like a smaller version of the Agni-V missile. One might expect two more tests to be conducted before the missile enters service. Of particular interest is the claim that the missile has a MIRV capability, with two such warheads being delivered.[101]

The Agni-III

At a glance, the Agni-III bears no physical resemblance to the Agni-II and is of a very different configuration, renouncing the tall, slender form for a shorter but wider arrangement. With three stages, the Agni-III represents what might be termed an intermediate stage of missile development in India, part way between the Agni-II and the definitive Agni-V. Once again using a rail-mobile TEL, the Agni-III represents the largest missile to date that India has fired from such a launcher.

After a failed test in 2006, the Agni-III completed the now normal series of three consecutive technical trials on 12 April 2007, 7 May 2008[102] and 7 February 2010. Thereafter, the SFC conducted user trials of the system on 21 September 2012[103], 23 December 2013 and 16 April 2015,[104] with the last one being on 27 April 2017.[105] A test at the end of 2019, however, ended in failure. Like the Agni-II,

Two photographs showing a test-launch of the Agni-Prime, on 28 June 2021. (DRDO)

An Agni-III missile during India's Republic Day Parade. Note the much bigger diameter of this design. (PIB)

a CEP in the 'two-digit' range (40m being suggested) was claimed. The Agni-III has a stated range of more than 3,500km, but it bears a striking resemblance to the Soviet-era SS-20 IRBM, leading to the possibility that its range could actually be as high as 5,000km.[106] While this has not been in any way verified or suggested further, the range of the Agni-III may indeed be somewhat understated.

While in chronological terms, the Agni-IV succeeded the Agni-III, it is suggested that the Agni-III represents an entirely new missile family, unrelated to the Agni-II and its successor the Agni-IV. Rather, it is said, the Agni-III represents the precursor to the Agni-V, which it closely resembles in size and configuration. It is of interest that upon an examination of the Agni-III TEL, there is a possibility that the vehicle is both road-mobile and rail-mobile. This continues a trend started with the Agni-I and Agni-II, which both possessed the ability to be launched from either road- or rail-mobile launchers, though in the case of the Agni-II, road-mobility seems to be debatable. Induction of the Agni-III reportedly began in 2011, but the unit designation and numbers inducted to date are unknown.[107]

The Agni-IV

With a range of 4,000km (or possibly more), the Agni-IV represents the successor to the Agni-II. Earlier known as the Agni-II Prime, the Agni-IV is no larger than its predecessor but has a significantly greater range and, being road-mobile, is even more flexible and survivable than the Agni-II. This reflects a greater use

The test-launch of the Agni-III from Wheeler Island (about 12km from Damra, off the Odisha coast) on 7 May 2008. (DRDO)

A detailed look at the Agni-III seconds after its launch. (DRDO)

of lighter composites and improved solid-fuel propellants, while improvements to the guidance system ensure that even at maximum range, the Agni-IV retains a 'two-digit' CEP. However, it is a streamlined and noticeably better-engineered system as compared to the Agni-II.

While its first test in December 2010 was a failure, technical trials on 15 November 2011, 19 September 2012[108] and 20 January 2014[109] – the latter in full user configuration – were deemed successful. Thereafter, the SFC commenced induction and carried out user trials on 2 December 2014[110] and 9 November 2015.[111] It is expected that the Agni-IV will supplement and then completely replace the Agni-II in production, and thus become the mainstay of the Indian IRBM force. It is also noteworthy that there have been persistent rumours of a canister-launched version of the Agni-IV being contemplated. Its most recent test on 2 January 2017 did not garner either the headlines or the attention of the 26 December 2016 test of the Agni-V. Nevertheless, from a practical and strategic standpoint, the Agni-IV represents India's most viable deterrent weapon against China until the Agni-V enters service.[112]

The Agni-IV entered service with the SFC in or around 2014, prior to its first user-associated trial in December that year. Since that time, it may be assumed that the Agni-IV is in production, and given its thus far trouble-free user-trials – three of which have now been conducted – it may also be inferred that the SFC is satisfied with the system and is increasingly comfortable with handling and operating the missile. This was confirmed on 23 December 2018, when the SFC conducted a successful launch of the Agni-IV.[113]

A test-launch of the Agni-IV missile. (DRDO)

The system, being both rail- and road-mobile, is even more flexible and survivable than the Agni-II. In fact, the Agni-IV is, by virtue of its weight and relative ease of transportation, a versatile delivery system which can be transported more easily than the larger (though much more capable) Agni-V.

The Agni-IV has been fired out to its full revealed range of 4,000km at least twice (19 September 2012 and 20 January 2014). It was fired out to 3,500km on 9 November 2015 and for about 3,000km on 15 November 2011 and 2 December 2014. With this demonstrated range, Chinese cities such as Beijing and Shanghai come within the range of the Agni-IV, even if the missile is sited in Uttar Pradesh or Bihar. The combination of road- and rail-mobility gives unprecedented flexibility to the Agni-IV.

There have been persistent rumours of a canister-launched version of the Agni-IV being contemplated. As with the Agni-V, a hermetically sealed canister offers the prospect of keeping the missile ready to fire, with a mated warhead. It is hoped that moves in this direction are taken with alacrity, with positive indications emerging (the author has seen requests for launcher and canister components), and that such a development will not be long in coming, increasing the already impressive flexibility of the Agni-IV.

With respect to warheads, the Agni-IV seems to be deployed with a fusion-boosted-fission weapon with a yield somewhere in the 150–500kt range. Needless to say, this figure is arrived at by some

The Agni-IV missile being launched just prior to delivery to the armed forces. (DRDO)

interpretation of comments, but the well-informed nature of these comments makes them impossible to ignore, as will be detailed later. It should be noted that the Agni-IV, with a relatively narrow nosecone, would probably not be a viable choice for MIRVs.

The Agni-V

While the first two launches of the Agni-V – on 19 April 2012[114] and 15 September 2013[115] – were of the standard Indian uncanisterised missile, following the pattern set with the Agni variants I through to IV, the tests on 31 January 2015[116] and 26 December 2016 were of a canisterised system mounted on a road-mobile platform. According to one report, by the usually well-informed Hemant Kumar Rout, the latter test was what might be termed a 'contained test fire' to a range of 2,500km, probably at a depressed or lofted trajectory to simulate a longer range.[117] However, some analysis of the NOTAM issued prior to the launch suggests a splashdown zone between 4,018km and 5,123km from the launch site. Given the fact that the Agni-V had already demonstrated a range of over 5,000km, this should not have been a cause for any concern.

However, what should continue to raise eyebrows is the persistent suggestion, backed by calculations based on flight time, speed and altitude of the Agni-V, that the range of the missile is in reality anywhere between 7,500km and 8,500km – as claimed by Chinese analyst Du Wenlong.[118] It would be fair to term the Agni-V an Intercontinental Ballistic Missile, because the range of 5,500km is the demarcation line between IRBMs and ICBMs in the normal use of the term, reflecting roughly the transatlantic distance between the Iberian Peninsula (or London) and the East Coast of the United States. Is India varying trajectories to simulate much longer ranges? This is a distinct possibility.

The Agni-V is, so far, the ultimate evolution of the Agni family. It combines the road-mobility of the Agni-I and Agni-IV with the configuration of the Agni-III but improves upon the latter by redesigning the missile and making great use of maraging steel and composite materials to make the rocket motor. Moreover, improvements in the guidance system reputedly brought down the CEP to 'single digits'. With these improvements in place, the Agni-V is now ready to enter production and induction into the SFC, which will then undertake user trials – starting with 'user-assisted' trials and then, as was the case with Agni variants I–IV, the SFC will take over subsequent trials.

In 2018, three tests of the Agni-V were conducted: on 18 January, 3 June and 10 December. On each occasion, the canister variant was tested by the SFC, in all probability indicating not only an IOC (Initial Operational Capability) but also the real likelihood that the production of the missile is in progress.[119]

In 2021, a story emerged of a pending test of the Agni-V to mark its induction in service – something that has *de facto* already happened. The usually reliable Indian defence journalist Hemant Kumar Rout made the following statement:

> Defence sources said the next trial of Agni-V missile assumes significance as it may be equipped with the MIRV capable of carrying multiple warheads. Though the MIRV capability of the missile was secretly tested during a multi-satellite launch, no live launch has been conducted so far. 'For the first time, the indigenous MIRV technology was tested successfully in [an] Agni P missile with the weapon delivering two manoeuvrable warheads at two separate locations. The MIRV capability of Agni-V will give India the much needed deterrence,' said the sources.[120]

This marks the first acknowledgement of a MIRV capability for the Agni-V and that India was validating this technology through its satellite launches as well as the recent Agni-P test.

The Agni-V is reportedly due to be followed by the Agni-VI, which would have an even greater range and multiple independently targetable re-entry vehicles (MIRVs).[121] In addition, a successor to the Agni-II and Agni-II is being designed, at the time of publication

An Agni-V missile in its uncanisterised version on a TEL. (PIB)

An Agni-V ballistic missile test-fired on 3 July 2018. Indian armed forces are regularly running such tests – not only to check functionality and reliability of missiles, but also to keep crews operating them current on all procedures. (DRDO)

Test-launch of an Agni-V ballistic missile on 10 December 2018 by the Strategic Forces Command. (DRDO)

termed the Agni-Prime, using technology from the Agni-IV.[122] It is hoped that these two missile projects will be pursued with alacrity and intensity.

What does the Agni saga tell us about the Indian ballistic missile programme? Unlike almost any other DRDO project, the IRBM/MRBM/ICBM programme has followed a path of constant product development and phased capability enhancement. This has meant that there is now a strong foundation in both design and systems engineering in respect of ballistic missiles, with fabrication and construction facilities being able to deliver materials and products of sufficient quality. There has been a steady path of technological

The Shaurya missile, for which renewed interest is being shown. (PIB)

achievement – replacing liquid-fuels with solid, improved guidance systems for accuracy and improved materials for lighter weights and greater ranges. While there have been delays and failures, the DRDO ballistic missile programme has done remarkably well on a limited budget.

Two other missiles are worthy of mention – the hypersonic Shaurya canister-launched missile – which was tested three times between 2008 and 2011 – and the tactical support missile Prahaar, which was tested in 2011. The Shaurya is purportedly a land-based version of the K-15 SLBM and offers a useful replacement for the Agni-I as it covers ranges from 700–1,800km.[123] The Prahaar is perhaps the ideal replacement for the Prithvi SS-150 – offering greater mobility, matching range and inherently greater flexibility.[124]

Despite the Shaurya and Prahaar offering distinct advantages over the existing Agni-I and Prithvi SS-150 missiles, it is unclear as to whether either missile has entered production, much less induction into India's armed forces. Following its 2011 test, there was some indication that the Shaurya was entering production, but this is as yet unclear. A follow-up test, indicating renewed interest in the missile, was conducted on 3 October 2020, at which point the missile was cleared for deployment.[125]

The future of the Indian land-based ballistic missile deterrent is likely to focus on increasing production and deployment of the Agni-IV and Agni-V missiles, with probable refinements to the accuracy of the two systems. Successors to the two missiles may take the form of being longer-ranged and, in the case of the Agni-IV, canister-launched. The development of a full-fledged ICBM is either already done with the Agni-V (given speculations over its true maximum range) or a question of priority. However, one may expect that a longer-range missile with MIRVs may be on the horizon in the relatively near future.

Numbers of Missiles

Two issues frequently escape discussion when dealing with India's missile arsenal – quantities and production capacity.

The strength of the Agni missile force is obviously a closely guarded secret. It has been suggested that the 334th and 335th Missile Groups, which operate the Agni-I and Agni-II missiles respectively, possess between eight and 12 TELs each. While it has been stated that another group has been raised on the Agni-III, its designation is unknown, and whether the TELs are accompanied by vehicles which carry reload missiles is equally unclear. It would be very odd if the total production of Agni missiles still remained in the low single digits per annum. It is in light of this that Dr S. Christopher's comments about 2015 production meeting only 20 percent of requirements is significant.[126] Though Dr Christopher went on to indicate that production of the Agni-I and Agni-II missiles would be increased, the fact is that there appears to be a significant disconnect between production and SFC requirements. As was noted earlier, as far back as 1998, directions were given to prepare production for up to 12 Agni missiles per annum.[127]

Quantities of missiles seem to still be a mystery. American assessments suggest in the cases of the Agni-II and Agni-III systems, fewer than 10 launchers are operational. This may be on the basis of early reports suggesting that each missile group was to have only eight launchers.[128] However, this may not be accurate as these numbers do not consider road-mobile variants of the Agni-II and Agni-III, nor do the numbers consider missiles and launchers that are kept in storage.

It has been suggested that at least two Agni-II missile trains are operational – perhaps with as many as eight launchers each.[129] Professor Bharat Karnad argues that from the Agni-III onwards, the systems would not be rail-deployed but stored in hardened mountain tunnel complexes. This would make considerable sense,

Table 13: Strategic Forces Command, Indian Army (Estimate for 2021)[133]

Unit	Base	Notes
334th Missile Group	Secunderabad	Agni-I, 8–16 launchers, 24–48 missiles
335th Missile Group		Agni-II, 8–16 launchers, 24–48 missiles
336th Missile Group		Agni-III, 8–16 launchers, 24–48 missiles
? Missile Group		Agni-IV, 16 launchers, 48 missiles
? Missile Group		Agni-V 16 launchers, 48 missiles
333rd, 444th & 555th Missile Groups		Prithvi SS-150, SS-250 & SS-350 – total of over 200 SRBMs deployed
? Missile Group		Shaurya missiles – very limited deployment

An indigenous TEL with four Nirbhay cruise missiles. (PIB)

to further reduce detection and also to enhance survivability.[130] It might be expected that ready-use and stored missiles would be so accommodated. Given the fact that at least two such complexes are already functioning, the conditions of storage and the maintenance of stored missiles must be examined on a continuous basis.[131]

To date there have been 12 Agni-II tests, more than a dozen Agni-I tests, nine Agni-III tests, eight Agni-IV tests and seven Agni-V tests. As noted earlier, India has had the habit of initiating the deployment of missiles after three successful technical trials. Extrapolating production from these tests is problematic.

If it is assumed that production commenced after the third successful technical trial, and that user trials would not involve testing more than one in five of any production batch, then it might be suggested that production of each type exceeds 20 missiles. This might also tally with the rough estimates available for the number of warheads held by India – if it is speculated that between 20 and two dozen of each of the Agni I–IV variants is in service to date, this would give India an eventually deployable strength of perhaps over 200 land-based missiles with some 80 launchers. Given that some estimates (which will be discussed in the following section) put the number of Indian nuclear warheads at between 110 and 130, the figure of 200 missiles and 80 launchers, plus aircraft-delivered weapons to be included in the overall total of deliverable warheads, may give an indication that the estimates of the Indian arsenal are somewhat understated. It is suggested that a realistic estimate of India's missile arsenal, as it emerges in the future, is going to be as listed here in Table 13.[132]

Sea-Based Missiles

In 2009, INS *Arihant* – the lead vessel of a class of at least three ballistic missile submarines (SSBN) – was launched. The vessel, which is still undergoing evaluation and sea trials, although reports emerged of its commissioning in August 2016, is an ambitious attempt to give India an operational and survivable nuclear triad.[134] Many Indian nuclear theorist and strategists have held that a force of SSBNs is essential for India to have a fully survivable nuclear deterrent, and the development of INS *Arihant* has largely been welcomed as a major step forward.

INS *Arihant* (S-2) is a relatively small SSBN, displacing around 6,000 tonnes and being powered by an 83MW pressurised light-water reactor with enriched uranium fuel. The vessel has a top submerged speed of 24 knots. It can carry 12 K-15 missiles or four K-4 missiles.

India is developing two SLBMs – the 750km-range K-15 and the 3,500km-range K-4 – to be deployed aboard the Arihant class of ballistic missile submarines. The K-15 underwent at least 12 development trials from a submerged pontoon aimed at simulating a submarine.[135] However, on 25 November 2015, an unarmed K-15 was purportedly fired from INS *Arihant*.[136] It should be stated that this has not yet been confirmed by Indian officials and no photographs have emerged of such a launch from the *Arihant*.

The K-4 was first tested on 24 March 2014 from a submerged pontoon,[137] and another pontoon launch followed on 7 March 2016.[138] Thereafter it was reportedly tested from INS *Arihant* itself on 31 March 2016.[139] This latter claim is probably untrue, instead referring to an ejection test from the *Arihant* rather than a full test. It would be surprising if such a major development were not highlighted in some way, but whatever the case, development of the K-4 is clearly well advanced. Four such missiles are to arm the Arihant class, and the K-4 bears a striking resemblance to the land-based Agni-III, both dimensionally and in respect of its performance.

The K-4 missile took some major steps towards deployment aboard INS *Arihant* when, on 19 January 2020, the first submerged launch from a pontoon was conducted to the missile's full range.[140] This was followed by a repeat, limited-range test on 24 January 2020.[141] At 12 metres in length, with a weight of 17 tonnes and a 3,500km range, the K-4 is a two-stage solid-fuel missile. In the aftermath of the second trial, it was reported that the missile was ready for induction and fitment aboard the *Arihant*.

Despite reports of commissioning and submerged K-15 and K-4 missile launches, it is suggested that some degree of caution should be exercised when claiming that an Indian sea-based deterrent is operational.

However, even armed with just the modest K-15 missile, INS *Arihant* has been commissioned. On 5 November 2018, it was disclosed that the vessel had completed a deterrent patrol of 20 days, ending on 4 November.[142] This at the very least indicates that the vessel has some operational capability and is considered to be deployable by the Indian SFC and Indian Navy.

It is only when other submarines of the Arihant class are commissioned that India will truly be deemed to have completed its sub-surface-based nuclear deterrent. Three more such submarines – one of which is the INS *Arighat* (S-3) – are in various stages of construction, and at the time of publication the launch of the latter may be expected soon.

Follow-on classes of SSBNs are expected to be slightly larger than INS *Arihant*, and therefore capable of carrying up to eight K-4 missiles. INS *Arighat* was launched in 2017, and it is possible that it might be commissioned in 2022, with two of an even larger variant – designated the S-4 and S-4* and armed with up to 12 K-4 missiles – being completed by 2024.[143] The completion of K-4 missile development may signal a renewed push for the completion of the Indian triad. There has been speculation that successor missiles to the K-4, called the K-6, with a 6,000km range, are in development, and that a larger submarine, called the S-5, of 13,000 tonnes displacement and with between 12 and 16 of the K-6 missiles, is the next step in India's SSBN programme.[144]

It may be in order to add a word here on the possibility of India deploying a surface ship-based deterrent. It is noted that at least two of the Indian Navy's very lightly armed Sukanya class of off-shore patrol vessels (OPVs) – INS *Subhadra* and INS *Suvarna* – have been fitted with storage, elevation and launching mechanisms for the Dhanush missile, with a 350km range.[145]

The Dhanush is a ship-launched version of the Prithvi surface-to-surface missile, and it is possible that its range could be enhanced by using a reduced payload warhead. However, while it is certainly conceivable that the Dhanush could be a viable nuclear delivery system, the Sukanya-class OPVs are a decidedly odd choice for an operational launch platform. They are very weakly armed, to the extent of being completely unable to defend themselves against even moderate levels of naval or air attack; they are best suited to the maritime constabulary role. Rather, it is suggested that the Dhanush system provided the Indian Navy and DRDO with an opportunity to develop elevation, stabilisation and launch mechanisms at sea. This may have aided in the development of both the K-15 and K-4 systems. Furthermore, the Dhanush system proved a useful target missile for India's ballistic missile defence system's early tests.

Summary

India's nuclear arsenal is still very much a matter of conjecture, with higher or lower estimates being given depending on the inclination of the analyst concerned. Slowly and steadily, India has moved towards the deployment of a nuclear triad, with land-, sea- and air-delivered components. In all three cases, however, the delivery systems are in a state of flux as new land-based missiles are introduced, aircraft await stand-off munitions and the first SSBN has yet to officially conduct a submerged firing of its main armament. India's cruise missile programme has been noticeably less fruitful, with consistent success proving elusive.

It should be noted that estimates of India's deployed missile arsenal are as yet relatively modest. It is possible that India could at present field somewhere in the vicinity of 200 MRBMs and IRBMs on 80 launchers, but any numbers significantly higher than that are purely speculative. While India's fissile stocks may allow for more weapons, delivery systems provide a limiting factor. Nonetheless, it can be concluded that the Indian arsenal is reasonably advanced and that a nuclear triad of increasing sophistication is rapidly emerging.

The only close-up photograph of INS *Arihant* released at the time of publication, taken while the submarine was underway for another test cruise, early after its service entry, in 2014. (PIB)

INS *Arihant* at sea, together with an Alouette III/Cheetah helicopter of the Indian Navy. (PIB)

The test-launch of a K-15 missile from submerged INS *Arihant*. (DRDO)

4

BALLISTIC MISSILE DEFENCES FOR INDIA – A NEW FACET TO INDIAN NUCLEAR STRATEGY[1]

Rather than view the Indian Ballistic Missile Defence (BMD) system as a separate and 'destabilising' component of India's force architecture, the BMD concept should be examined within the wider and more important parameters of the Indian strategic air defence network and its gradual modernisation and expansion to deal with a wider variety of targets. If looked at in this context, India's BMD efforts are much more of an attempt to improve the range of targets that can be engaged by its air defence network.

It should be noted that the present Indian air defence network is undergoing a substantial upgrade process. Unfortunately, the current air defence network has a large number of obsolete systems, which will mean that the upgrade process will be a time-consuming one. Yet the gradual upgrading of the air defence network will also facilitate the integration of the BMD system when it becomes available. It must also be said that the present state of Indian indigenous BMD efforts have been somewhat exaggerated. While the programme has achieved tangible successes, its exo-atmospheric component is some way from entering service and India will probably field an imported system to cater for exo-atmospheric interception for some time to come.

BMD systems in the Indian scenario have to be seen in the context of the upgradation of the IAF's air defence network to cater for new technologies and more potent threats.

Strategic Air Defences in India

India, with its vast airspace, maintains an advanced Air Defence Ground Environment System (ADGES). This system, along with the civilian air traffic control, is responsible for the detection, identification and, if necessary, the interception of aircraft in Indian airspace. The air defence network is also in the process of being upgraded to cater for ballistic missile threats. Before examining the system in detail, a quick overview is in order. India's air defence network is essentially divided into two parts – the ADGES and the Base Air Defence Zones. These two components are closely linked and share information relating to air defence tasks. The ADGES consists of an array of radars along India's western and northern borders, as well as a network of mobile systems in the north-east and south of the country. Southern India is still peculiarly ill-served by air-defence sensors.

The ADGES network is responsible for overall airspace management and detection of intruders. It also controls and coordinates the air defences for large area targets. The Base Air Defence Zones (BADZ), meanwhile, as the name implies, are tasked with the defence of high-value targets such as air bases, nuclear installations and key military installations. The BADZ is a scaled-down ADGES network, limited to an arc of 100km, compared to hundreds of kilometres for the ADGES. The BADZ is a far more concentrated air-defence environment than the ADGES and provides the only gap-free air defence cover in most sectors. In addition to these networks, India is now establishing an anti-tactical ballistic missile screen, with new radars and weapons. It is not clear whether this will be incorporated into the BADZ or whether it will comprise a separate network. This Anti-Tactical Ballistic Missile (ATBM) screen is slowly taking shape and news of its structure is still awaited. It is noteworthy that India's traditional concern has been its vulnerability to low-level aerial attacks in areas without adequate radar coverage. This is slowly changing to reflect the multiplicity of threats at all altitudes which can emerge.

Indian Air Defences: Sensor Network

The Indian Air Defence Ground Environment System employs a three-tier detection network. While this system is currently in the process of a major modernisation programme, the basic structure of the ADGES network will remain unchanged. The first layer, rather surprisingly, consists of Mobile Observation Posts (MOPs). These remain among the most reliable of the early-warning mechanisms available to the Indian Air Force. The MOPs consists of two-man teams equipped with a HF/VHF radio set and field glasses. The personnel in the MOP are very well versed in the visual identification of aircraft as well as their general direction of flight. The MOPs are scattered along the borders at random intervals, ranging between 25km and 45km. The MOPs usually give the first warning of airborne intrusion, the general direction of the attack and, more often than not, the number of aircraft and their type. The MOPs are assisted in this task by personnel from the Indian police forces and Railway Protection Force, who are given some training in aircraft identification. These agencies report via a communications system based on both HF/VHF radio sets as well as telephone lines. A more advanced communications system based on fibre-optic cables and satellite communications is also available to assist the MOPs in reporting to the radar picket line.

The radar picket line, which lies about 150km behind the MOPs, consists of a number of radar clusters. These comprise three radar stations separated at a distance of the sum of their radii. The equipment issued to these clusters generally comprises one license-made Soviet ST-68/U and two P-18/-19 radars. These are then flanked by two P-12/-15 radars. The ST-68/U acts as the Control and Reporting Centre (CRC). This may have now changed somewhat, as the ST-68/U, which was plagued with some nagging development problems, has largely replaced older Soviet-made equipment. Moreover, India has been license-producing the French-designed TRS-2215D 3-D surveillance radar for a number of years and has derived an indigenously built radar from it, the PSM-33 Mk. 2. These have probably supplanted most of the older Soviet-bloc equipment. It should be pointed out that these radars are all long-range surveillance types, with ranges in excess of 300km and good performance against targets flying at all altitudes – even those employing electronic countermeasures (ECM). These radar pickets are responsible for giving accurate information on the intruding force to the air defence control centres (ADCC) located behind the radar picket line. The picket line and the ADCC are separated by a first layer of air defence weapons which are the first to engage the intruders.[2]

The backbone of the Indian ADGES network is the THD-1955 3-D long-range surveillance radar. This radar, originally of French design, has been license-produced in India for a number of years.

A THD-1955 3D long-range surveillance radar of French origin, seen installed in an early warning radar site in India. (IAF)

Though by now somewhat elderly, it still has sterling performance characteristics and is capable of maximum detection ranges of up to 1,000km, though in peacetime the Indian Air Force usually limits its power to a 400km detection range. These form the core of the ADCCs. The ADCCs also keep in touch with the BADZ control centres.

Like the ADGES, the BADZ consists of three layers. The first of these are the Mobile Observation Posts, followed by a mixed layer of weapons and their associated radars along with a picket line of low-level radars. These are in turn supported by anti-aircraft artillery batteries. This network is controlled by a ST-68/U radar. The BADZ provides comprehensive and gap-free coverage over its assigned area of responsibility. Some observers have likened the BADZ set-up to the defence pattern of a carrier battle group. Any aircraft attacking a vital military target, therefore, not only has to get past the ADGES, but also the far more formidable BADZ. This has serious implications for the attacking force.

SAM Assets

SAM units comprise 30 squadrons of SA-3b Pechoras and eight squadrons of SA-8b OSA-AKM systems, deployed to protect key air bases as well as some major military/industrial centres. Though the SAMs are old, they have been updated periodically and, when operating as part of the BADZ, are deployed in such a manner as to minimise their shortcomings. In addition, a large number of L-40/70 radar-directed 40mm anti-aircraft guns and man-portable Igla-1M SAMs are deployed to provide a 'last-ditch' tier of 'hard-kill' defences. It should be pointed out, however, that this system is geared up to the defence of point targets and not for overall area defence. It also lacks a viable capability against ballistic missiles.

Upgrading the System

With this in mind, the Indian Air Force has begun a substantial modernisation of its strategic air defences, but while progress has undoubtedly been made, strategic SAM defences remain weak. The IAF is in the process of inducting seven additional squadrons of Akash SAMs to augment the eight squadrons delivered to date, and has begun the upgrade of eight of the Pechora squadrons with an option of upgrading five more.[3] In addition, at least nine squadrons of a 70km-range MRSAM are being procured, with the possibility that several more squadrons may follow, which, in conjunction with future Akash procurements, could replace the entire Pechora force.[4] Replacement of the OSA-AKM is in progress, with four squadrons of SPYDER SAMs being delivered and a new Quick Reaction SAM under development. Further development of the Akash SAM – the Akash-1 Prime and the Akash NG – is underway, with the former being optimised for deployment at very high altitudes while the latter is a completely different system with a new solid-fuel pulse rocket propulsion and an AESA radar integrated into a self-contained launcher complex. The missile itself has an enhanced range of between 40km and 70km, as opposed to the Akash-1/ Akash-1 Prime range of just 25–30km. Trials of both the Akash-1 Prime and Akash NG commenced in 2021 and have so far been successful, with the Akash-1 Prime missiles seeming likely to be ordered to augment reloads for existing IAF Akash launchers.

For long-range SAMS, the IAF has chosen to procure five squadrons of the S-400 SAM, with an effective range of 400km and the additional capability to offer significant interception capability against ballistic missiles.[5]

Radar upgrades

The process of replacing the legacy radars of the Indian Air Force is slowly making progress. A critical element in this is the Arudhra version of the Elta M-2084 radar. Orders for 34 radars of this type

A test-launch of an Akash surface-to-air missile (SAM), developed by India's state-owned DRDO. The missile is designed to reach a maximum speed of about Mach 3.5. (DRDO)

were placed in 2009, and between direct supply and production of an Indian version, this radar will become the backbone of the new air defence network. The Arudhra will act as an early warning and target engagement radar as opposed to the airspace surveillance/ air traffic management radars, which currently dominate the IAF's inventory. The IAF will be able to employ the Arudhra EL/M-2084 to provide early warning of inbound tactical ballistic missiles and cruise missiles, whether air-launched, ground-launched or perhaps even launched from surface or subsurface assets. Installation of these radars will initially take place in and around Jamnagar, Mumbai and the National Capital Region.[6]

Technically speaking, the AESA-based EL/M-2084 represents a quantum leap in technology for the IAF – especially compared to the PSM.33 and TRS-2215D radars that it will replace. It merges all land-based radar functions – weapons location, air surveillance and air defence functions, inclusive of cruise and tactical ballistic missiles. The EL/M-2084 is able to detect and track incoming targets and, in the case of missiles and other such projectiles, can calculate the anticipated impact and launching points, subsequently passing on target data to the relevant air-defence weapons systems. As an initial step, this system would enable India to detect and potentially engage incoming tactical ballistic missiles but would not be able to cope with longer-range systems.[7]

The sensor network will be completed with a series of low-level and medium-level radars. Low-level surveillance is to be catered for by a total of 67 low-level air-transportable radars (LLTR), including 19 180km-range, three-dimensional THALES-built Ground Smarter GS-100 radars which were ordered in November 2009. Six of these were supplied directly, with the remaining 13 being license-assembled by HAL. Each radar will be accompanied by operational and communication shelters, an energy sub-system, a mobility sub-system and personnel living quarters – making them virtually self-contained units. These will augment the 34 active phased-array EL/M-2084 medium-power radars (MPR) detailed above.[8]

In addition, a series of Indian-developed systems will include the DRDO-developed and Bharat Electronics Ltd (BEL)-built S-band Aslesha three-dimensional micro-radar, the Army-specific Bharani man-portable radar, and 30 (20 more to be ordered) 180km-range Rohini S-band central acquisition radars. The Aslesha, which weighs 250kg, uses low-probability-of-intercept frequencies to look out for terrain-hugging tactical UAVs (Unmanned Aerial Vehicles) over mountainous terrain out to 50km. The IAF has to date ordered 21 of them, and first deliveries took place in January 2008. In contrast, the Bharani is a two-dimensional L-band gap-filler system now in series-production for the Indian Army. It has a range of 40km and can track up to 100 airborne targets. To date, 16 Bharanis – meant to be used in conjunction with VSHORADS/ MANPADS – have been ordered and deliveries are in progress. In addition, also under delivery are 29 THALES Reporter tactical control radars for the Army's upgraded ZU-23 and L70/40 air defence guns.[9]

The IAF has a total of three A-50 Phalcon AEW (airborne early wearning) platforms, supplemented by three ERJ-145-based Netra AEW systems developed by DRDO. Six more systems based on Airbus A-320 airframes acquired from Air India have been cleared for procurement, and conversion of the aircraft has begun The IAF is now gearing up to induct new-generation S-band long-range surveillance radars (LRSR) and is also hoping to acquire an additional nine ELTA Systems-built L-band EL/M-2083 'Airstar' aerostat-mounted high-power radars (HPR) to add to the two already in service, but to date no progress has been reported in this regard. For the LRSR requirement, a competition is presently underway between the ELTA Systems-built EL/M-2288 AD-STAR, THALES-built Ground Master 400 and SELEX Sistemi Integrati's RAT-31SL.[10] Twenty Ground-Master 400 systems were also ordered in 2009 for the IAF.[11]

These new radar acquisitions will be integrated with the IAF's existing 32 new mobile control and reporting centres (MCRC), 12 air defence control centres (ADCC), 24 air defence direction centres (ADDC) and some 40 terminal weapons control centres (TWCC) along India's western and north-eastern borders. This will enable the slow but progressive replacement of the abovementioned existing ST-68U gap-filler radars and related 19ZH6 command-and-control consoles, and P-18/NRS-12 and P-19 gap-filler radars. It is anticipated that the LRSR will replace the THD-1955 radars, but this is some way from happening. In the interim, new inductions will enable the replacement of the P-30/NRS-20, P-37 and P-40 gap-filler/target engagement radars, and THALES-built TRS-2215D and BEL-built PSM-33 Mk2 airspace surveillance radars, all of which were inducted in the 1970s and 1980s.[12]

Both the Indian Army and Navy are desirous of obtaining aerostat-mounted EL M-2083 radars. In the Army's case, the intention is to acquire six such systems to enable the Corps of Army Air Defence to detect and track both ballistic missiles and terrain-hugging cruise missiles launched from Pakistan, while the Indian Navy is reportedly asking for two EL/M-2083s. Capability wise, the 1,700kg EL/M-2083 'Airstar' is mounted inside a 240ft-long aerostat that is perched at altitudes of up to 4,000ft, and uses electronically steered multi-beam techniques to detect terrain hugging airborne targets – combat aircraft, helicopters, cruise missiles and UAVs –

An ELTA ELM-2084 radar. Originally designed to support the Spyder-MR SAM-system, this Israeli-made ground-based mobile 3D multi-mission radar is capable of detecting and tracking both aircraft and ballistic missiles and providing fire-control guidance for surface-to-air missiles. (IAF)

at ranges of up to 300km, while the trajectories of ballistic missiles can be accurately plotted up to 500km away. There has been a renewed emphasis on acquiring these systems on the part of all the three services, despite some earlier delays in additional purchases of the type.

These systems would enhance India's TBM (tactical ballistic missile) defences and cruise missile defences but would offer little by way of assistance in the detection of longer-ranged ballistic missiles, which would require radars of considerably greater capability and which have yet to enter service with the Indian armed forces. Yet to see India's BMD efforts as something separate from its overall air defence modernisation programme is to miss the very essence of that programme, which seeks to completely transform the Indian air defence network from a 1980s vintage system to one that is state of the art. This is somewhat different from the approach taken in the United States, where the BMD system is distinct from the limited air defence system available for the defence of that country owing to a significantly reduced aerial threat. In contrast, the IAF must cater for manned aircraft, short- and long-range cruise missiles, as well as tactical and strategic ballistic missiles. The IACCCS (Integrated Air Command, Control and Communications System), as described below, will be an important component of establishing defences against these threats.

The Integrated Air Command, Control and Communications System.

Phase 1 of the Indian Air Force's Integrated Air Command, Control and Communications System, which seeks to provide a layered, hardened and in-depth air defence command, control and communications network, achieved full operational capability by June 2012 with the operationalisation of the AFNET digital information grid, which is an IAF-owned, operated and managed fully secure and reliable network. IACCCS operations will use AFNET, enabling the integration of all ground-based and airborne sensors, air defence weapon systems and command and control nodes, allowing a coordinated and efficient response to any aerial threat from any sector.[13]

The Indian Air Force has also sought to enhance its airspace management and surveillance capabilities to cater for both peacetime and wartime exigencies. To this end, aviation expert Prasun Sengupta states, the IAF has initiated a multi-phase $1.3 billion programme under which a state-of-the-art joint civil/military sub-continental airspace control system is being developed using the following fundamentals: unity of effort, common procedures and simplicity. The IAF's terminal area air traffic services and airfield management expertise and en route airspace/air corridor management are also being upgraded. It is hoped that the final result will be the creation of a vastly expanded air defence identification zone (ADIZ) and provision of a real-time recognised air picture (RAP). The upgraded ADIZ will extend the IAF's airspace management and surveillance coverage (using ground-based sensors) up to 500 nautical miles away from India's territorial boundaries. When fully implemented, new-generation ATCR-33S and SIR-S primary/secondary surveillance radars and their related joint air traffic control and reporting centres (JATCRC) will be operational at IAF air bases in Adampur, Agra, Ambala, Bagdogra, Bareilly, Bhatinda, Bhuj, Bidar, Chabua, Chandigarh, Gorakhpur, Gwalior, Halwara, Hashimara, Hindon, Jaisalmer, Jamnagar, Jodhpur, Jorhat, Kalaikunda, Nal, Naliya, Pathankot, Pune, Sirsa, Suratgarh, Tezpur, Uttarlai, Yelahanka and Zopuitlang in Lunglei district in southern Mizoram.

The IACCCS can be further integrated with the Indian Army and Navy radar networks and with the civilian airspace surveillance system, and to enable the provision of an integrated air situation picture to operators to carry out an AD (air defence) role. It is the intention of the IAF that all of its assets – air, ground and, eventually, space – would be connected to create total situational awareness of a region. For the first time, the IAF will be close to achieving a complete level of integrated coverage of all of India's airspace. This will not only dramatically enhance basic air defence operations but will provide an essential first step on which a BMD detection, tracking and engagement system is to be built. To this end, the IACCCS will also coordinate the early-warning and response and interception aspects of India's planned layered, ground-based, two-tier ballistic missile defence network that is now in the development stages, and which depends heavily on the IACCCS.

Communications for the IACCCS is provided through the fibre-optic network-based AFNET, which replaces the IAF's troposcatter-based communications network. This system was developed at a cost of Rs10.77 billion in collaboration with the US-based Cisco Systems Inc., HCL Infosystems Ltd and Bharat Sanchar Nigam Ltd (BSNL). It incorporates the latest traffic transportation technology in the form of internet protocol (IP) packets over the network, using multi-protocol label switching (MPLS). A large voice-over-internet-protocol (VoIP) layer with stringent quality of service enforcement will facilitate robust, high-quality voice, video and conferencing solutions. Having established these critical components, the IACCCS can now be integrated with a large number of new-generation

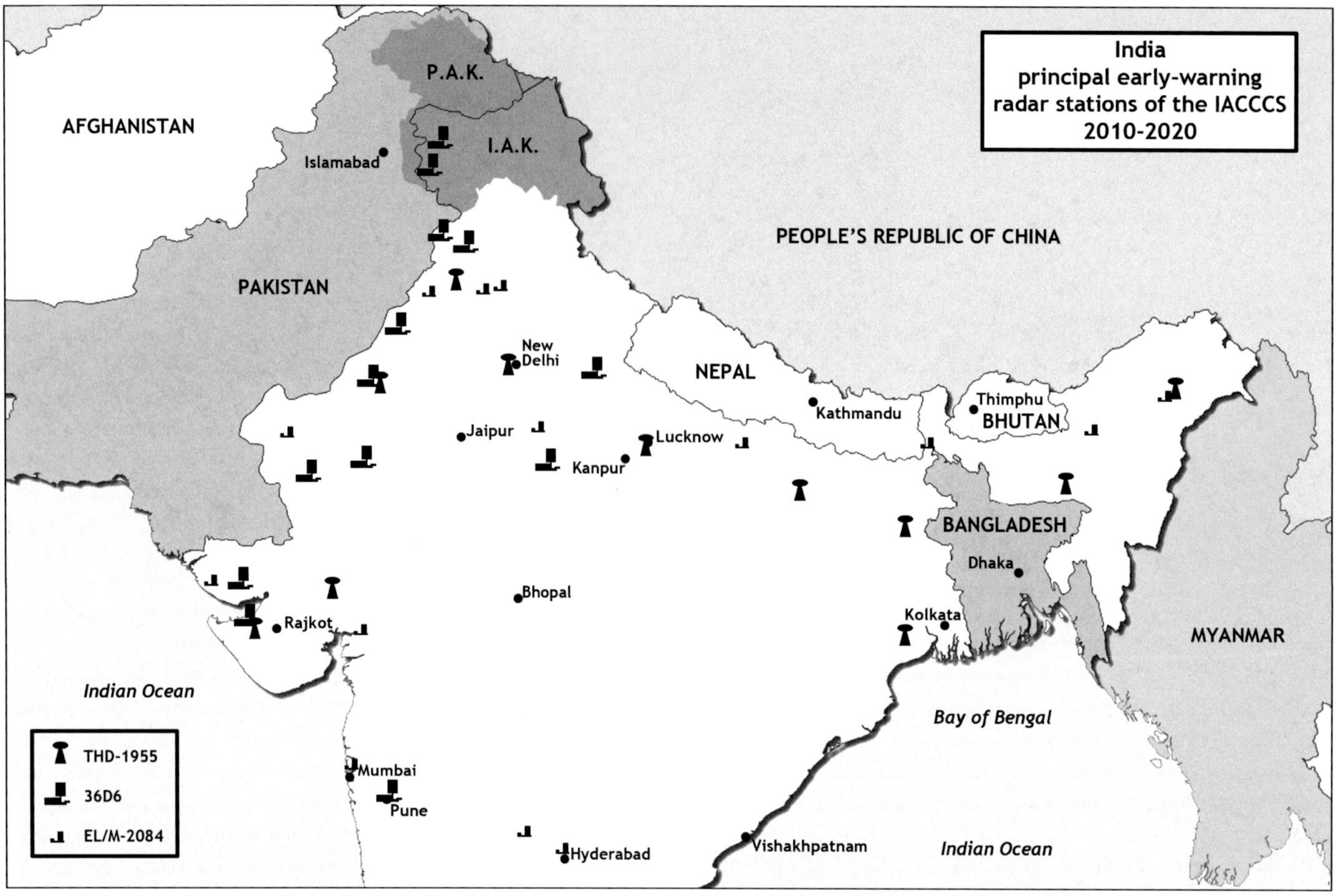

A map of the approximate positions of principal early-warning radar stations of the IACCCS as of 2010–2020. During this decade, the IACCCS radar network was significantly improved. Major radar stations equipped with French-made THD-1955s and Russian 36D6s are now supported by indigenous Rohini and Israeli-made ELTA EL/M-2084 3D radars, while low-altitude coverage has been improved through the acquisition of 19 Thales BM-100s. The mass of modern radars is mobile: in the case of a war, most of them would frequently change their positions. (Map by Tom Cooper)

ground-based radars that are now in the process of being delivered or are on order. These radars will be able to deal with airspace surveillance in search of airborne targets (such as manned aircraft, ballistic and cruise missiles, attack helicopters and unmanned aerial vehicles), as well as coastal surveillance or ground surveillance. This would enable India to mount a coordinated response to an attack from any direction and through any medium.[14]

In addition to ground-based sensors and radars, the IAF is increasingly making use of India's space assets. India's employment of satellites for military purposes has been gradually revealed. In 2016, half-a-dozen ISRO satellites were used to obtain ground information for the surgical strike carried out by the Indian Army against terrorist targets in Pakistan-occupied Kashmir (PoK). Prior to this, the Indian Navy used the GSAT-7 satellite to assist it in search and rescue operations and was able to seamlessly network some 60 ships and 75 aircraft during the 2014 Theatre-level Readiness and Operational Exercise conducted in the Bay of Bengal.[15]

Shortly after the 2017 launch of the CARTOSAT-2E satellite, it was revealed that some 13 satellites were being used for surveillance purposes. This number is set to grow as newer versions of the IRS series begin to enter service. However, barring a few examples, none are dedicated military satellites, being used for both civilian and military purposes.[16]

India's National Technical Research Organisation (NTRO), which is controlled by the Research and Analysis Wing, India's premier intelligence agency, makes extensive use of data from India's IRS, RISAT and CARTOSAT satellites to aid it in building a comprehensive intelligence picture. One may expect, as the Indian space programme grows in capability – and as the military defines its needs more clearly – space-based radars and missile-launch detection sensors to become a priority alongside the more mundane requirements of imagery and communications.

The Role of the IACCCS in the Indian BMD Network

Dealing with manned aircraft and cruise missiles is well within the technological capacity of India's existing systems, which have undergone periodic upgrades as needed. However, the need for a BMD system – encompassing both exo-atmospheric and endo-atmospheric interception assets – presents a completely different challenge and has presented the IAF and DRDO with a number of technological and practical constraints. At the outset, it has to be reiterated that to date, India has not deployed any BMD systems and its capability is still nascent, though steady progress has been made.

At the current rate of progress, it is unlikely that the initial components of such a two-tier BMD network, comprising both endo-atmospheric and exo-atmospheric missile interceptors, are to be commissioned before 2022.[17] It is probable that the initial Indian BMD system will be composed of a mixture of imported and domestic BMD interceptors and radars.

At present, for fire-control purposes the BMD system uses ELTA Systems-built EL/M-2080 'Green Pine' ground-based active phased-array L-band long-range tracking radar (LRTR). Two of these radars were supplied in late 2001 under the US$50 million 'Project Sword Fish' to the DRDO by the ELTA Systems Group subsidiary of Israel Aerospace Industries.[18] This has led to some confusion, as some have assumed the name 'Sword Fish' to be a DRDO-developed radar. India did indeed write three million lines of software code for the Battle Management/Command, Control, Communications & Intelligence (BM/C³I) centre, the hub of software and hardware systems.

Furthermore, transmission links to the interceptor missile based on jam-proof CDMA technology with multiple data transmission links have been set up so that if one is jammed the others could function. In addition, Israeli inputs were sought and obtained for designing and fabricating the BM/C³I centre, which not only acts as the DRDO's primary BMD engagement simulator, but is also being used for evolving BM/C³I concepts, for defining BMD goals and developing BMD doctrine, for evaluating candidate systems architectures, for serving as the principal prototyping-cum-validation tool for the BMD's BM/C³I algorithms, and for defining the human role in the BMD battle. The BMD's endo-atmospheric element makes use of the THALES Raytheon-supplied S-band Master-A engagement radar.[19]

Ballistic Missile Defences for India: Interceptor Missiles

India's efforts to obtain a ballistic missile interception capability date back to between 1996 and 1998. News began leaking out about the deployment from 1998 onwards of an Anti-tactical Ballistic Missile screen. This system was to comprise the Russian S-300V ATBM (SA-12) and India's own 'Akash' missile, which has a considerable ATBM capability. In March 1997, the Indian press confirmed these reports, stating that one S-300V squadron was being purchased, with more to come in the future. These reports proved to be incorrect, and as far as is known no variant of the S-300 has entered Indian service. The S-300 saga illustrates some of the many pitfalls of Indian defence reporting and coverage, where apparently definitive reports of acquisitions turn out to be completely false. This has the dual effect of confusing analysis while simultaneously decreasing the credibility of analysts who, in good faith, believe such media reports.

India's quest for a BMD system has had many false starts, and while still very much a work in progress has made some tangible steps towards the development of a viable system. Early efforts focused on evaluating the potential of the Akash SAM to intercept short-range ballistic missiles. To date, this remains a potential development but has never been demonstrated. Other efforts included stillborn attempts to acquire S-300 PMU-1 and S-300V systems from Russia, as noted above, and Arrow-2 systems from Israel.

Beginning in late 2021, the first of five S-400 squadrons were due to be delivered to the IAF, and these will have a significant ATBM role.[20]

India's indigenous efforts at developing a BMD system have had the misfortune of being plagued by a series of misstatements by DRDO officials which claimed deployment readiness when it was evident that the Indian system was more of a proof-of-concept technology demonstration until later versions of its endo- and exo-atmospheric interceptors became available.[21]

C4ISR

DRDO's overall BMD C4ISR architecture is intended to consist of both over the horizon and X-band fire-control radars which detect and track incoming missiles, plus a mission control centre (MCC) that fuses input (which may also come from satellite-based sensors), processes it and then sends orders for engagement to launch control centres (LCCs) situated up to 1,000km away via mobile communication terminals (MCTs).

Repeated tests of the two-tier system, including the latest PDV (Prithvi Defence Vehicle) test, have given enough confidence to the DRDO to recommend the freezing of the current configuration for Phase 1. Both the radars and the LCCs receive and send information via target update transmitters (TUTs) based on CDMA technology. The MCTs of the MCC are themselves connected via an IP wide area network, while datalinks for the entire setup also include fibre-optic communication channels and line of sight relays.[22]

A deployed Indian BMD system would consist of several launch vehicles, radars, LCCs and the MCC. Using a secure communication system, the MCC and perhaps secondary sites would be used to

An AAD during its first test. (DRDO)

An AAD being launched from its mobile launcher. (DRDO)

link a geographically widespread network. The MCC is a software intensive component which receives information from radars and satellites which is then processed by 10 simultaneously running computers, connected to all other elements of the system through a WAN (wide area network). These would enable the MCC to perform tasks such as target classification, target assignment and interception success assessment deciding the number of interceptors required for the target for an assured kill. After performing all these functions, the MCC assigns the target to the LCC of a launch battery. The LCC then computes the necessary time required and optimal window for launching the interceptor, based upon information received from a radar based on the speed, altitude and flight path of the target.[23]

Endo-Atmospheric Interceptors – the AAD

The Advanced Air Defence (AAD) system is an anti-ballistic missile designed to intercept incoming ballistic missiles in the endo-atmosphere at altitudes of between 15km and 30km. It is a single-stage, solid-fuelled missile. Its guidance is based on an inertial navigation system, with midcourse updates from ground-based radar, and it has an active radar homing in the terminal phase. It is 7.5m (25ft) tall, weighs around 1.2 tons and has a diameter of less than 0.5m (1ft 8-in.).[24]

Compared to the exo-atmospheric systems, the AAD has been tested often and against relatively realistic targets. On 6 December 2007, the AAD successfully intercepted a modified Prithvi-II missile acting as an incoming ballistic missile enemy target. The endo-atmospheric interception was carried out at an altitude of 15km. The AAD was launched when the Prithvi reached an apogee of 110km. The AAD, with the help of midcourse updates and its terminal seeker, manoeuvred itself towards the target and scored a hit on it at a speed of Mach 4. Ground- and ship-based radars detected the formation of a large number of tracks, signifying that the target had broken into multiple pieces. This was confirmed by thermal cameras located on Wheeler Island. This was followed on 26 July 2010 by the AAD being successfully test-fired from the Integrated Test Range (ITR) at Wheeler Island.

On 6 March 2011, India launched its indigenously developed interceptor missile from the Odisha coast. India successfully test-fired its interceptor missile, which destroyed a 'hostile' target ballistic missile, a modified Prithvi, at an altitude of 16km over the Bay of Bengal. The interceptor, an AAD missile positioned at Wheeler Island, about 70km across the sea from Chandipur, received signals from tracking radars installed along the coastline and travelled through the sky at a speed of Mach 4.5 to destroy it. The interceptor missile had its own mobile launcher, a secure data link for interception, independent tracking and homing capabilities and sophisticated radars.[25] Another test was conducted on 10 February 2012.

Finally, on 23 November 2012, India again successfully test-fired its home-made supersonic AAD interceptor missile from a defence base off the coast of Odisha. The test-firing was part of India's efforts to create a missile defence shield against incoming enemy missiles. The AAD interceptor missile was again fired from Wheeler Island and successfully destroyed in mid-air an incoming ballistic missile launched from the Integrated Test Range in Chandipur.[26]

The Definitive Endo-atmospheric Interceptor: Upgrading the AAD

On 22 November 2015, the upgraded AAD was successfully tested. The anti-ballistic missile took off at 9:40 a.m. from Wheeler Island soon after it received the command to waylay and destroy an incoming electronically simulated target missile. Conditions similar to the launch of a target missile from Balasore were simulated electronically, and upon receiving its coordinates the interceptor missile, travelling at supersonic speed, engaged and destroyed the 'virtual target' in mid-flight.[27] The first test against a live target took place on 15 May 2016. During this test, the DRDO officially reported that the upgraded AAD interceptor intercepted and destroyed a Prithvi ballistic missile fired from a ship.[28] However, some reports suggest that the intercept failed, with the upgraded AAD failing to launch, much less intercept the target.[29]

A further test took place on 1 March 2017, when the upgraded AAD interceptor achieved a terminal intercept of the target missile at an altitude of 15km. Another test on 28 December 2017 achieved a hit-to-kill intercept at an altitude of 15km.[30]

The upgraded AAD configuration for these tests was similar to what would be offered for production and deployment. This test achieved both an actual intercept of an incoming ballistic missile target as well as an explosive intercept, as the pre-fragmented warhead on-board the upgraded AAD was detonated using the missile's radio proximity fuse (RPF).[31] The upgraded AAD was initially guided by its on-board inertial navigation system (INS), which received continuous updates about the incoming target's trajectory from ground-based radars through a secure data link. Subsequently, a radio frequency seeker in the upgraded AAD's nose-cone section tracked the target while an intercept course was plotted by its on-board computer.[32]

To summarise, as compared to the exo-atmospheric interceptors, development of the AAD and the upgraded AAD endo-atmospheric interceptor missiles has witnessed greater urgency. The AAD was test-fired on 6 December 2007, 6 March 2009, 15 March 2010, 26 July 2010, 6 March 2011, 10 February 2012 and 23 November 2012. As has been detailed above, upgraded AAD missile test-firings commenced on 6 April 2015, and were followed by further test-firings on 23 November 2015, 15 May 2016, 1 March 2017 and 28 December 2017.

In the future, it is expected that the upgraded AAD will be supplemented with the AD-1 and SA-2. Both the upgraded AAD and AD-1 are able to sustain up to 30G, which makes them unstable at an altitude of over 35km.[33] Fire-control cues are provided by the DRDO-developed, 400km-range Arudhra S-band active phased-array medium-power radar, for which orders worth Rs 8 billion have already been placed with Larsen & Toubro, while TATA Power's Strategic Electronics Division has been contracted for the upgraded AAD's TEL vehicle.[34]

The version of the LRTR used in the May 2017 upgraded AAD mission is an L-band array that can track a ballistic target with a radar cross-section (RCS) of 0.1sqm from over 1,500km away. The MFCR (Multi-function Fire Control Radar), which is an S-band array, has a tracking range of over 370km for a target with an RCS of 0.3sqm. These detection figures may be sufficient to detect a re-entry vehicle, but this is debatable.[35]

Exo-Atmospheric Interceptors

India's efforts to develop an exo-atmospheric interceptor have been markedly slower than those made in respect of its endo-atmospheric interceptors. This has made any claim of either Phase 1 or Phase 2 of India's BMD system being ready for deployment somewhat incredible. To date, only four tests of exo-atmospheric interceptors have been carried out, two of which were with what can best be described as a proof-of-concept demonstrator, based on the liquid-fuelled Prithvi SSM. This is an impractical system for deployment as an interceptor missile, and as such India has opted to move away from that towards a potentially more efficient and effective system. A critical aspect for successful exo-atmospheric interception is target detection. Sufficient warning time must be available for a successful intercept. India is yet to develop or deploy a missile early-warning radar suitable for the detection of longer-ranged missiles and lacks space-based radars. Until such time as these systems are developed and deployed, India's exo-atmospheric interception capability will remain somewhat limited.

India's requirements for an exo-atmospheric interceptor have acquired a degree of urgency, with the Indian Air Force identifying multiple targets which need to be protected. To this end, as will be discussed later, India and Russia have been negotiating for the sale of S-400 TMD systems to India specifically for the purpose of providing a degree of protection for vital areas and potential targets within India. The procurement of the S-400 would provide India with a considerable degree of exo-atmospheric interception capability, as one version of the system is capable of intercepts at altitudes of up to 185km, approximating the performance of the US THAAD system.

India's efforts have not been insubstantial in this regard, with intercepts achieved at altitudes of 47km, 75km and most recently 97km, with the existing PDV missile purportedly having the capability to intercept targets at altitudes exceeding 150km and itself being the precursor for a second phase of development aimed at producing an interceptor with an intercept ceiling of 300km. Yet India's progress in this respect has been slow. Only four tests have been conducted of exo-atmospheric interceptor missiles over an 11-year period. This could be due to the greater technological challenges faced in the development of these missiles, or it could also be due to the greater priority assigned to the AAD and upgraded AAD endo-atmospheric interceptor missiles.

The Prithvi Air Defence Missile

The first test of India's indigenous BMD was the test of the PAD (Prithvi Air Defence) missile in November 2006. During this exercise, a PAD missile successfully intercepted a modified Prithvi-II missile at an altitude of 47km. The Prithvi-II ballistic missile was modified successfully to mimic the trajectory of M-11 missiles.

A second test took place in March 2009, when India conducted a test using a ship-launched Dhanush missile (the naval version of the Prithvi missile) as the target, simulating a missile with a range of 1,500km. Sword Fish (LRTR) radar was used for tracking, and it was destroyed using a PAD missile at an altitude of 75km.[36]

The PDV

For exo-atmospheric interceptions, the PAD has been replaced by the much more capable Prithvi Defence Vehicle missile, which is intended to carry out interceptions up to an altitude of 150km. On 27 April 2014, the first PDV was successfully tested against an electronic target. This simulation seemed to be intended to validate the parameters of the interceptor missile rather than to achieve an actual interception, and as such has to be viewed as a test launch rather than an actual engagement.[37] Furthermore, it was used to

The Prithvi Air Defence missile. (DRDO)

validate the interceptor's integration with the detection, tracking and automated launch control systems.

This was followed by an actual intercept on 11 February 2017, when a target missile was engaged in a successful hit-to-kill intercept at an altitude of 97km. While this is a considerable achievement, unlike earlier instances when claims of near-deployment were being made, the DRDO issued a much more nuanced statement indicating the requirement for more tests and a more effective detection system, though the capability of the PDV was undeniable.[38]

By achieving an exo-atmospheric intercept at an altitude of 97km, the 11 February test validated, among other things, an improved guidance algorithm used for this test-mission, as the incoming missile target had deviated significantly from what would have allowed an intercept along a typical minimum energy trajectory (MET).[39]

Compared to the PAD, which was essentially a high endo-atmospheric system with a maximum ceiling of around 85km, the PDV is a genuine exo-atmospheric interceptor capable of destroying targets at altitudes of up to 150km. With only a couple of tests having been completed, however, the PDV is some distance from production or deployment.

ASAT or Enhanced BMD Capabilities?

Into any discussion of the PDV must come its latest incarnation – the Indian ASAT system which is called the PDV Mk. II.

On 27 March 2019, India announced that it had intercepted a satellite – later identified as an Indian Microsat-R – at an altitude of 300km with a ground-based missile. This at one stroke dramatically enhanced India's capabilities in the spheres of both anti-satellite defence and ballistic missile defence.

The project – code named XSV-1 – led to the fielding of a missile known as the PDV Mk. II which has at least some traits in common with the PDV exo-atmospheric interceptor of the Indian BMD programme.

The PDV during its first test. (DRDO)

The PDV has demonstrated an intercept altitude of 97km. (DRDO)

The PDV Mk. II clearly builds on the PDV and is a three-stage missile. While details are as yet unclear, the upper stage clearly derives from the PDV but its lower stages are much more powerful to allow for the increased performance necessary to reach an altitude of 300km.

The PDV Mk. II looks to be an amalgam of PDV kill technology combined with the booster stage of a ballistic missile – probably the submarine-launched K-4 ballistic missile. The missile has a kinetic energy warhead which does not explode, but which destroys its target by sheer velocity and mass. This requires the system to have a hit-to-kill capability as a near miss will not suffice to destroy the target. This was successfully demonstrated during the 29 March intercept and speaks highly of the accuracy and precision of the intercept.

The intercept target and its altitude of 300km was selected to minimise debris from the intercept. However, the chairman of the DRDO, G. Sateesh Reddy, has indicated that the PDV Mk. II can intercept all satellites up to an altitude of 1,000km. This clearly places India's capabilities close to those demonstrated by China, Russia and the United States.

Deployment of a BMD System is Getting Closer

In pursuance of a BMD network, the first steps have been taken to secure clearance for the deployment of a limited BMD system around Delhi, with the IAF and DRDO approaching the government for such a system, which will be undertaken as part of a broader, integrated national air defence network which will fully integrate Indian Army and Navy radars and SAMs with the IAF's network.[40]

5

CONCLUSION – INDIA'S EVOLVING NUCLEAR ARSENAL AND STRATEGY[1]

India is perhaps unique among nuclear powers in that its doctrine, strategy and capability remain discussed under a veil of secrecy so dense that making informed assessments of Indian intentions, force levels, capabilities or even threat perception is somewhat difficult. Nevertheless, the country has moved towards a coherent doctrine that reflects a degree of strategic maturity, but which has been poorly articulated by successive governments of India. Furthermore, there is a degree of uncertainty regarding India's capabilities which, despite the attempts of successive governments and their scientific advisors to clarify, remains to this date.

India's path to nuclear weaponisation, initially hesitant, has intensified towards the development of a viable triad, with air-, land- and sea-based systems. India's arsenal seems to be evolving, with a combination of improvements to existing missile systems as well as the development of a new series of submarine-launched missiles. Questions remain as to whether India will begin development of ICBMs or more advanced cruise missiles for its aircraft, but there are technical obstacles to the development of any such new systems.

How will India's nuclear strategy evolve? Will India opt for a review of its nuclear no-first-use doctrine or will it choose to manifest its nuclear strategy with NFU as its central tenet? The following observations might be made in respect to the challenges India faces:

1. India will find itself faced with growing Pakistani and Chinese nuclear arsenals, which will include increasingly advanced IRBMs and MRBMs – possibly with MIRVs.
2. Pakistan's development and deployment of tactical nuclear weapons has effectively lowered the nuclear threshold.
3. China's nuclear doctrine and strategy are continuously evolving as that nation continues its military modernisation efforts.

To date, India's response to these challenges is a phased and evolutionary approach to the development of its delivery systems, an assertion of its stated nuclear doctrine and a guarded approach to revealing its levels of preparedness to deal with a nuclear strike. India has further consistently reiterated its self-imposed moratorium on nuclear testing and has tried to refute persistent attempts to question the quality of its nuclear deterrent and the size of its arsenal. Moreover, despite separating its civilian and nuclear programmes, it has kept a number of nuclear reactors outside the ambit of IAEA safeguards but has never admitted their use for weapons. From these broad parameters, we may draw the following conclusions about existing nuclear strategy:

1. India's nuclear arsenal is evolving, with the Agni-IV and Agni-V missiles destined to form the core of its land-based mobile IRBM/ICBM force. Its efforts to develop a nuclear triad are at present in progress, with INS *Arihant* undergoing trials while its sister ships are being prepared for launch. The Indian Air Force's role has been somewhat downgraded as its aircraft currently lack a stand-off long-range cruise missile capable of delivering nuclear warheads. The SFC is the designated body tasked with the management of the Indian arsenal, though the delivery systems remain part of the armed service to which they were assigned. As one of only two operational tri-service commands in India – the other being the Andaman and Nicobar Fortress Command – the SFC has devised mechanisms to ensure the smooth operational deployment of the components of the Indian arsenal.
2. While not overtly stating so, India's strategy against Pakistan has been largely decoupled from that against China. This is evident from the priorities accorded to weapons development – the Agni-I MRBM remains India's primary weapons delivery system against Pakistan, replacing the shorter-ranged Prithvi missiles, and little priority has been accorded to the Shaurya canister-launched MRBM that might have been a possible candidate to supplement and later supplant the Agni-I. In contrast, India has given priority to developing the Agni-II, Agni-III, Agni-IV and Agni-V systems, all of which are likely to be deployable against China as opposed to Pakistan. India has sought to enhance the readiness levels of its SFC through periodic tests of its ballistic missiles by the respective units operating the Prithvi and Agni families of missiles but

has been very wary of revealing any details of the SFC, its functioning or its order of battle.

3. The size of the Indian arsenal is clouded in mystery. Part of that is understandable ambiguity, but much of the confusion stems from the status of the eight unsafeguarded CANDU-type power reactors that could be used for the production of weapons-grade plutonium. Estimates of the Indian arsenal appear only to consider production from the Dhruva reactor and the now-decommissioned CIRUS research reactor. Moreover, Indian nuclear weapons design will have a direct impact on the size of a deployable arsenal. Uncertainty over India's thermonuclear weapons persists, with speculation that the large payload capacity of the Agni series of missiles is intended for fusion-boosted-fission weapons with yields of several hundred kilotons, but heavier than comparable pure fusion weapons and significantly more profligate in the use of fissile material. Miniaturisation does not appear to be an issue.
4. Similar confusion persists over the number of nuclear delivery systems currently deployed by India. There is some evidence to show the production of over 200 Prithvi SRBMs, but little evidence to suggest that production on a similar scale has taken place for the Agni missiles. Circumstantial evidence suggests one group is being created for each Agni variant, with an establishment of 12 launchers and an anticipated two missiles per launcher. Evidence also points to some 48 Agni-V missiles being planned for production over a four-year period, with orders for that number of trailers being placed in 2015. It is entirely unclear whether similar numbers are planned for the other Agni variants, although the number of deployable missiles does place an upper limit on the number of nuclear weapons India has likely deployed or is likely to deploy in the short to medium term. Whether or not the nascent SSBN and SLBM force is deployed as expected – with delays being inevitable – there will be additional force accretions to the missile and warhead stockpiles. Furthermore, as India moves to deploy canister-launched missiles – the Agni-V initially – the concept of having a de-mated arsenal for recessed deterrence has been seemingly abandoned in favour of a more practical approach but one which will require additional safeguards.
5. India does not appear to have developed a nuclear war-fighting doctrine for its armed forces but has taken steps to improve its CBRN passive defences, especially in its Army formations facing Pakistan. While some limited publicity has been accorded to CBRN-protected troops participating in exercises, there is as yet no indication of such passive defensive measures being extended to troops facing China, although it is unlikely that any provision of such equipment will be publicised. Little indication of either IAF or Indian Navy CBRN defensive measures has been given to date. While much research work has been done on BMD systems, a deployable two-tier BMD network is some years away.
6. Indian command, control and communications systems are reputed to be hardened to survive a nuclear attack and the effects thereof. This has been publicly stated and is applicable to the Nuclear Command Authority and the secondary NCA as well. However, it is somewhat unclear as to just how survivable the Indian arsenal is. India has emphasised mobile land-based missile systems, with submarine-launched missiles now also being developed. These are somewhat more survivable than fixed silos, but it has been noted that India has invested in survivable mountain storage sites for its missiles and warheads to safeguard them from an enemy first-strike while maintaining the ability to be rapidly deployable. As India moves towards a viable SSBN capability – especially when longer-range SLBMs become available – the survivability quotient of the Indian arsenal will be significantly enhanced, with no adversary assured of being able to launch an effective counter-force first-strike on India.
7. India's nuclear doctrine is stated to be based on the principles of no-first-use and 'massive retaliation'. Despite calls for the doctrine to be reviewed and revised in light of Pakistan's deployment of tactical nuclear weapons, India has remained adamant that it will adhere to the no-first-use principle and continues to state that any nuclear first-strike would lead to massive retaliation. However, it is to be noted that there is considerable divergence of opinion as to whether the Indian doctrine is inflexible or whether it possesses a degree of nuance and flexibility that could perhaps indicate that India may choose to respond 'disproportionately' rather than 'massively', with the connotations that the latter policy entails for solely attacking counter-value targets. This debate continues, and it is interesting to note that India has not chosen to officially clarify its position in respect of its doctrinal nuances.

An enhanced PDV – the PDV Mk. II – was used to demonstrate an ASAT capability in 2019. (DRDO)

In light of what is currently known about India's strategy, it is pertinent to note that, whether intentionally or not, India has effectively created a somewhat ambiguous nuclear strategy. This has some benefits as it keeps adversaries guessing as to its intent. On the other hand, it also proves to be somewhat confusing as to India's intentions and its capabilities in the event of a nuclear or other CBRN attack from an adversary. This will be interesting in the event of an attack by a non-state actor against an Indian civilian target.

To date, India has not shifted towards an overt nuclear war-fighting doctrine for the battlefield. Its own tactical nuclear weapons capability is existent mainly as a potential, while that of Pakistan – its likely battlefield adversary – is increasingly overt. However, India's doctrine seems to have a degree of flexibility, which will allow for a relatively flexible response to the use of nuclear weapons on a battlefield. It could be that any doctrinal changes will see efforts made to bring some clarity to India's response in this regard without veering away from its current doctrine of no first-use of nuclear weapons, at least not to any appreciable extent.

The Indian arsenal, and with it, India's nuclear strategy, will continue to grow and evolve, supported by an astonishing degree of political consensus among India's political parties. India has taken a long and somewhat convoluted route to becoming a nuclear weapons state. Decades of ambiguity ended in 1998, and in the two decades after that the Indian government has steadily increased its nuclear weapons potential without overtly commissioning dedicated weapons reactors and improved its delivery systems.

What the future holds is unclear but will be interesting and possibly will involve new families of ballistic missiles – both land- and sea-based – while its doctrine will continue to evolve at a pace commensurate to its perceptions and capabilities. What remains to be seen is whether India moves towards capabilities beyond its ability to financially sustain without compromising either its economy or its conventional forces. To date, India has been very careful in avoiding this trap and has done so quite successfully. Yet the lure of larger arsenals and more assertive nuclear doctrines can be irresistible, and while India has avoided going down this path to date, the challenges it faces from its two main adversaries – Pakistan and China – could compel it to pursue a more vigorous deterrent posture, with attendant changes to the size of the Indian arsenal and the Indian nuclear doctrine.

BIBLIOGRAPHY

Articles

'Agni III could have 5,000 km range: Russian General', *PTI* article carried in *The Hindu*, http://www.defencetalk.com/forums/missiles-wmds/indian-nuclear-missile-development-news-discussions-7241-11/

'Agni-III successfully test fired from Odisha coast', *Economic Times* (16 April 2015), http://economictimes.indiatimes.com/news/defence/agni-iii-successfully-test-fired-from-odisha-coast/articleshow/46941664.cms

'Agni-V can reach targets 8,000 km away: Chinese expert', *The Hindu* (21 April 2012), http://www.thehindu.com/todays-paper/tp-in-school/agniv-can-reach-targets-8000-km-away-chinese-expert/article3337202.ece

'Ballistic and Cruise Missile Threat 2017', *Defense Intelligence Ballistic Missile Analysis Committee*, http://www.nasic.af.mil/Portals/19/images/Fact%20Sheet%20Images/2017%20Ballistic%20and%20Cruise%20Missile%20Threat_Final_small.pdf?ver=2017-07-21-083234-343, p.25

'BARC scientists defend estimates of n-tests', *The Hindu* (19 September 1998)

'CABINET COMMITTEE ON SECURITY REVIEWS PROGRESS IN OPERATIONALIZING INDIA'S NUCLEAR DOCTRINE', http://pib.nic.in/archieve/lreleng/lyr2003/rjan2003/04012003/r040120033.html

'Defence Acquisition Council okays purchase of Russian air defence missile systems', *Press Trust of India* (17 December 2015), http://www.dnaindia.com/india/report-defence-acquisition-council-okays-purchase-of-russian-air-defence-missile-systems-2156916

'Draft Indian Nuclear Doctrine', *Arms Control.org*, https://www.armscontrol.org/act/1999_07-08/ffja99

'DRDO told to test Upgraded Agni in August', *Deccan Chronicle* (15 July 1998)

'Electronic Weapons: Instant Radar', *Strategy Page* (21 February 2009), https://www.strategypage.com/htmw/htecm/articles/20090221.aspx 'Government approves appointment of 3 experts as NDMA members', *Press Trust of India* (21 December 2014), http://www.dnaindia.com/india/report-government-approves-appointment-of-3-experts-as-ndma-members-2045959 (accessed 10 November 2017)

'Government to restructure and revamp NDMA', *GK Today* (30 September 2014), https://currentaffairs.gktoday.in/government-restructure-revamp-ndma-09201414966.html (accessed 10 November 2017)

'India acquires Green Pine radars from Israel', *Press Trust of India* (28 June 2002), https://timesofindia.indiatimes.com/india/India-acquires-Green-Pine-radars-from-Israel/articleshow/14351441.cms

'India building nuclear-proof bunkers for top leaders', *Silicon India* (23 September 2003), http://www.siliconindia.com/shownews/India_building_nuclearproof_bunkers_for_top_leaders-nid-20887-cid-Top.html

'India capable of building nuke deterrence up to 200 kilotons: Kakodkar', *Deccan Herald* (24 September 2009), http://www.deccanherald.com/content/27047/india-capable-building-nuke-

'India conducts first night trial of Agni-II missile', *India Today* (16 November 2019), https://www.indiatoday.in/india/story/india-conducts-first-night-trial-of-agni-ii-missile-1619704-2019-11-16

'India goes Thermonuclear', http://www.fas.org, p.2

'India has fissile material for 2,000 warheads: Pak media', *Times of India* (10 September 2015), http://timesofindia.indiatimes.com/world/pakistan/India-has-fissile-material-for-2000-warheads-Pak-media/articleshow/48895568.cms

'India Plans Next Generation Bombproof Shelter for 108 Fighter Jets' *Sputnik News* (3 July 2017), https://sputniknews.com/military/201707031055187413-india-shelter-jets/ (accessed 4 September 2017)

'India successfully test-fires 3,500 km nuclear-capable missile K-4', *New Indian Express* (20 January 2020), https://www.newindianexpress.com/states/odisha/2020/jan/20/india-successfully-test-fires-3500-km-nuclear-capable-missile-k-4-2091838.html

'India successfully test-fires AAD missile interceptor', *NDTV.com* (23 November 2012)

'India Successfully Test-fires N-capable Agni-I Ballistic Missile', *News 18* (22 November 2016), http://www.news18.com/news/india/india-successfully-tests-nuclear-capable-agni-i-ballistic-missile-1314397.html

'India Successfully Test-Fires New Interceptor Missile', *Outlook India* (27 April 2014), https://web.archive.org/web/20140428163936/http://news.outlookindia.com/items.aspx?artid=838755

'India successfully test-fires nuclear capable Agni-II missile off Odisha coast', *News 18* (9 November 2014), http://www.news18.com/news/india/india-successfully-test-fires-nuclear-capable-agni-ii-missile-off-odisha-coast-724952.html

'India successfully test-fires supersonic interceptor missile', *Tribune India* (15 May 2016), http://www.tribuneindia.com/news/nation/india-successfully-test-fires-supersonic-interceptor-missile/237107.html

'India successfully tests nuclear capable Agni IV missile', *The Hindu* (3 December 2014), http://www.thehindu.com/todays-paper/tp-in-school/india-successfully-tests-nuclear-capable-agni-iv-missile/article6656285.ece

'India test fires medium range nuclear capable Agni-II missile', *Economic Times* (20 February 2018), https://economictimes.indiatimes.com/news/defence/india-test-fires-agni-ii-missile-off-odisha-coast/articleshow/62993975.cms?utm_source=contentofinterest&utm_medium=text&utm_campaign=cppst

'India test fires nuclear capable Agni-5 missile, 2nd test in six months', *Economic Times* (10 December 2018), https://economictimes.indiatimes.com/news/defence/india-test-fires-nuclear-capable-agni-5-missile-2nd-test-in-six-months/articleshow/67023684.cms

'India test-fires Agni V with range as far as China', *Hindustan Times* (16 September 2013), http://www.hindustantimes.com/india/india-test-fires-agni-v-with-range-as-far-as-china/story-28IHgfrhxGgUt9XLNPRiwN.html

'India test-fires nuclear-capable Agni III ballistic missile', *Economic Times* (27 April 2017), http://economictimes.indiatimes.com/news/defence/india-test-fires-nuclear-capable-agni-iii-ballistic-missile/articleshow/58396268.cms

'India test-fires nuclear-capable Nirbhay cruise missile', *Times of India* (17 October 2014), http://timesofindia.indiatimes.com/india/India-test-fires-nuclear-capable-Nirbhay-cruise- missile/articleshow/44845526.cms

'India to join ICBM club soon – Interview with Dr. S. Christopher',*NDTV* (11 July 2015), http://www.ndtv.com/video/news/news/india-to-join-icbm-club-soon-374683

'India to test fire Agni-V by year-end', *The Hindu* (3 June 2011), http://www.thehindu.com/sci-tech/science/India-to-test-fire-Agni-V-by-year-end/article13821309.ece

'India's Nuclear Tests: News reports and "eyewitness" accounts', https://seismo.berkeley.edu/~rallen/research/nuke/India.May98/news.reports.html

'India's nuclear triad is complete with INS Arihant ending its first deterrence patrol', *The Hindu* (5 November 2018), https://www.thehindu.com/news/national/ins-arihant-completes-deterrence-patrol-india-declares-nuclear-traid-operational/article25425436.ece

'Indian army may soon get bio-chem suits', *Rediff.com* (11 May 2011), http://www.rediff.com/news/report/indian-army-may-soon-get-bio-chem-suits/20110511.htm

'Indian Navy successfully test fires Dhanush missile: All you need to know', *India Today* (26 November 2015), http://indiatoday.intoday.in/education/story/dhanush/1/531950.html

'Interceptor missile scores "direct hit"', *The Hindu* (7 December 2007), http://www.hindu.com/2007/12/07/stories/2007120761241800.htm

'Interceptor missile test fired successfully', *NDTV.com* (6 March 2011)

'Long range strategic missile Agni-IV test-fired', *The Hindu* (19 September 2012), http://www.thehindu.com/sci-tech/science/long-range-strategic-missile-agniiv-testfired/article3914340.ece

'Missile defence shield to be ready in three years: India', *Dawn* (13 December 2007), https://www.dawn.com/news/280120

'National Disaster Management Plan Part 1: Draft', p.61

'Nirbhay, India's Indigenous Cruise Missile, Fails Test Midway', *NDTV* (16 October 2015), http://www.ndtv.com/india-news/nirbhay-indias-indigenous-cruise-missile-fails-midway-1233086

'Now, India has a nuclear triad', *The Hindu* (18 October 2016), http://www.thehindu.com/news/national/Now-India-has-a-nuclear-triad/article16074127.ece

'NUCLEAR ANXIETY; Indian's Letter to Clinton On the Nuclear Testing'. *New York Times* (13 May 1998), https://www.nytimes.com/1998/05/13/world/nuclear-anxiety-indian-s-letter-to-clinton-on-the-nuclear-testing.html?auth=login-facebook

'Nuclear Capable Agni-IV missile successfully test fired', *Economic Times* (23 December 2018), https://economictimes.indiatimes.com/news/defence/nuclear-capable-agni-iv-missile-successfully-test-fired/articleshow/67215177.cms?from=mdr

'Odisha: Nuclear capable Agni-II missile successfully test fired', *IBN Live* (7 April 2013), https://archive.is/20130628175643/http://www.bharat-rakshak.com/NEWS/newsrf.php

'Operation Shakti', *Nuclear Weapons Archive – India's Nuclear Weapons Program*, https://nuclearweaponarchive.org/India/IndiaShakti.html

'Pakistan was to deploy nukes against India during Kargil war', https://m.economictimes.com/news/defence/pakistan-was-to-deploy-nukes-against-india-during-kargil-war/articleshow/50019153.cms

'Press Statement by Dr. Anil Kakodkar and Dr. R. Chidambaram on Pokhran-II tests', *Press Information Bureau* (24 September 2009), http://pib.nic.in/newsite/PrintRelease.aspx?relid=52814 (accessed 20 September 2017); see also 'Pokhran-II tests were fully successful; given India capability to build nuclear deterrence: Dr. Kakodkar and Dr. Chidambaram, *Press Information Bureau* (24 September 2009), http://pib.nic.in/newsite/PrintRelease.aspx?relid=52813

'Prithvi missiles to be replaced by more-capable Prahar: DRDO', *The Hindu Business Line* (30 June 2013), http://www.thehindubusinessline.com/news/prithvi-missiles-to-be-replaced-by-morecapable-prahar-drdo/article4866081.ece

'Shadow of an Indian H-Bomb', *Foreign Report* (13 December 1984), p.1

'Smiling Buddha', *Nuclear Weapons Archive*, http://nuclearweaponarchive.org/India/IndiaSmiling.html.

'South Asian Nuclear-Weapon-Free Zone', *The Arms Control Reporter* (1993), p.454. B.176

'Sub-sonic cruise missile "Nirbhay" successfully test-fired', *The Hindu* (16 April 2019), https://www.thehindu.com/news/national/other-states/sub-sonic-cruise-missile-nirbhay-successfully-test-fired/article26849032.ece

'The Long Pause 1974-1989', *Nuclear Weapon archive*, http://nuclearweaponarchive.org/India/IndiaPause.html

'Third Pakistani nuclear reactor operational: US magazine', *The Nation* (7 July 2014)
http://nation.com.pk/islamabad/07-Jul-2014/third-pakistani-nuclear-reactor-operational-us-magazine.

'Use plural, India has thermonuclear bombs: Kakodkar', *CNN-IBN Interview Transcript* (13 December 2009)

'We Have Low-yield N-weapons to Ward Off India's War Threat: Pakistan', *India Today* (20 October 2015), http://indiatoday.intoday.in/articlePrint.jsp?aid=503185

'Week view No. PR94/2011-ISPR' (19 April 2011), https://www.ispr.gov.pk/front/main.asp?o=t-press_release&date=2011/4/19

Abbas, Shahid K., 'Brahmos anti-ship missile to be produced in 2003', *Rediff.com* (1 June 2002), http://www.rediff.com/news/2002/jun/01war11.htm

Ahmed, Ali, 'The Need for Clarity in India's Nuclear Doctrine', *Institute for Defence Studies and Analyses* (11 November 2008) https://idsa.in/idsastrategiccomments/TheNeedForClarityInIndiaSNuclearDoctrine_AAhmed_111108 (accessed 13 November 2008)

Ahmed, Ali, 'Tit for Tat: A Nuclear Retaliation Alternative', *Institute for Defence Studies and Analyses* (3 October 2011), https://idsa.in/idsacomments/TitforTatANuclearRetaliationAlternative031011

Ahmed, Mansoor, 'India's Nuclear Exceptionalism, Fissile Materials, Fuel Cycles, and Safeguards' Harvard Kennedy School, Belfer Centre for Science and International Affairs: Discussion Paper (May 2017)

Albright, D. 'The Shots Heard "Round the World"', *The Bulletin of the Atomic Scientists* (July/August 1998)

Albright, D. & Kelleher-Vergantini, 'India's Stocks of Civil and Military Plutonium and Highly Enriched Uranium', *IPFM*, Global Fissile Material Report (2015)

Allard, Tom & Da Costa, Agustinus Beo, 'Exclusive: Indonesian militants planned "dirty bomb" attack – sources', *Reuters* (25 August 2017), https://www.reuters.com/article/us-indonesia-security/exclusive-indonesian-militants-planned-dirty-bomb-attack-sources-idUSKCN1B51FW

Arms Control, 'India's Draft Nuclear Doctrine', https://www.armscontrol.org/act/1999_07-08/ffja99

Aroor, Shiv, '10 Reasons Why the Indian Rafale is Evolution Itself', *Daily O* (4 July 2017), http://www.dailyo.in/variety/rafale-aircraft-brahmos-nuclear-defence/story/1/18157.html (accessed 4 September 2017)

Aroor, Shiv, 'True BrahMos Unleashed Today, Next 900km Weapon', *Livefist Defence*, https://www.livefistdefence.com/2017/03/true-brahmos-unleashed-today-next-1000-km-weapon.html

Aroor, Shiv, 'India deploys long-range missile Nirbhay to counter Chinese threat at LAC', *India Today* (28 September 2020), https://www.indiatoday.in/india/story/india-deploys-long-range-missile-nirbhay-to-counter-chinese-threat-at-lac-1726255-2020-09-28

Badri-Maharaj, Sanjay, 'Civil Defence Capabilities of the Indian State', *Bharat Rakshak Monitor*, vol. 4, No. 2 (September–October 2001)

Badri-Maharaj, Sanjay, 'The IAF Fleet Needs Rejuvenating, And Needs It Now', *Swarajya* (9 January 2017), https://swarajyamag.com/defence/the-iaf-fleet-needs-rejuvenating-and-needs-it-now

Badri-Maharaj, Sanjay, 'The Importance of Passive and Active CBRN Defensive Measures', *Institute for Defence Studies and Analyses: IDSA Issue Brief* (17 October 2016), https://idsa.in/issuebrief/importance-of-passive-and-active-cbrn-defensive-measures_sbmaharaj_171016

Badri-Maharaj, Sanjay, "The Indian Air Force's Declining Squadron Strength – Options and Challenges', *Institute for Defence Studies and Analyses – Issue Brief* (3 November 2017), https://idsa.in/issuebrief/the-indian-air-force-declining-squadron-strength_sbmaharaj_031117

Balachandran, G. & Patil, K., 'Revisiting India's Nuclear Doctrine', *Institute for Defence Studies and Analyses* (27 March 2017), https://idsa.in/idsacomments/revisiting-india-nuclear-doctrine_gbala-kpatil_270317

Balachandran, G. & Patil, Kapil, 'Revisiting India's Nuclear Doctrine', *Institute for Defence Studies and Analyses* (27 March 2017), https://idsa.in/idsacomments/revisiting-india-nuclear-doctrine_gbala-kpatil_270317

Baranwal, Jayant, *SP's Military Yearbook 1992–93* (New Delhi: Guide Publications, 1993)

Barbosa, Rui C., 'China continues Build Up of Yaogan-30 constellation', *Nasa Spaceflight* (26 December 2017), https://www.nasaspaceflight.com/2017/12/china-continues-yaogan-30-build-up/

Barnaby, Frank, 'Trials provide data for range of weapon yields', *Jane's Defence Weekly* (27 May 1998)

Basu, Nayanima, 'Armed Forces Undergoing Nuclear, Biological Warfare Training', *The Hindu Business Line* (11 August 2017), http://www.thehindubusinessline.com/news/national/armed-forces-undergoing-nuclear-biological-warfare training/article9813830.ece

Bhaduri, S., 'The Artillery Division – Part II', *Indian Defence Review* (April 1992)

Bhalla, Abhishek & Bagga, Bhuvan, 'The Uttarakhand Tragedy Casts Shame on India's Disaster Management. So Why DID the Committee meant to Plan for Flood Emergencies Fail to Meet for FOUR YEARS?', *Daily Mail India* (24 June 2013), http://www.dailymail.co.uk/indiahome/indianews/article-2347624/NDMA-The-Uttarakhand-tragedy-casts-shame-Indias-disaster-management-So-DID-committee-meant-plan-flood-emergencies-fail-meet-FOUR-YEARS.html

Bhaskar, Chitrapu Uday, 'India's Nuclear Deterrence Strategy', *The Hindu* (30 March 2010), http://www.thehindu.com/books/Indias-nuclear-deterrence-strategy/article16630532.ece

Bhatia, Vimal, 'IAF to Deploy New Radar near Border Areas', *Times of India* (2 October 2011), https://timesofindia.indiatimes.com/city/jaipur/IAF-to-deploy-new-radar-near-border-areas/articleshow/10204373.cms

Carnegie Endowment for International Peace, 'A Conversation With Gen. Khalid Kidwai, 2015', *Carnegie International Nuclear Policy Conference 2015*, transcript 4–5 (23 March 2015), http://carnegieendowment.org/files/03-230315carnegieKIDWAI.pdf

Chan, Minnie, 'China Plans Sea-Based Anti-Missile Shields "for Asia-Pacific and Indian Ocean"', *South China Morning Post* (8 February 2018), https://www.cnbc.com/2018/02/08/china-plans-sea-based-anti-missile-shields-for-asia-pacific-and-indian-ocean.html

Chandra, Satish, 'Revisiting India's Nuclear Doctrine: Is it Necessary?', *Institute for Defence Studies and Analyses* (30 April 2014), https://idsa.in/issuebrief/RevisitingIndiasNuclearDoctrine_schandra_300414

Chellaney, Brahma, 'Nuclear Deterrent Posture', in Brahma Chellaney (ed.), *Securing India's Future in the New Millennium* (New Delhi: Orient Longman Limited, 1999), pp.209–214

Chengappa, Raj, 'The Missile Man', *India Today* (15 April 1994), pp.40–42.

Chinese State Council Information Office, 'China's National Defense in 2008', http://www.china.org.cn/government/central_government/2009-01/20/content_17155577_9.html

Chopra, Air Marshal Anil, 'India's Military Space Program', *South Asia Defence and Strategic Review*, http://www.defstrat.com/india%E2%80%99s-military-space-program

CISF, 'CISF Fire Wing', http://www.cisf.gov.in/fire-wing/

Civil Defence, 'Fire Cell', http://ndrfandcd.gov.in/fire.aspx

Conrad, Edward E. *et. al.*, 'Collateral Damage to Satellites from an EMP Attack', *Defence Threat Reduction Agency* (August 2010), http://www.dtic.mil/dtic/tr/fulltext/u2/a531197.pdf

Cordesman, Anthony H., with Kendall, Joseph & Colley, Steven, 'China's Nuclear Forces and Weapons of Mass Destruction', *CSIS*, working draft (20 July 2016)

DDR Staff, 'DRDO's AAD Ballistic Missile Defence Interceptor Heads Toward Induction With Latest Test', *Delhi Defence Review* (29 December 2017), http://www.delhidefencereview.com/2017/12/29/drdos-aad-ballistic-missile-defence-interceptor-heads-toward-induction-with-latest-test/

Disaster Management Act (23 December 2005), No. 53.

Dubey, A.K., 'Bhopal Gas Tragedy: 92% Injuries Termed "Minor"', *First14 News* (26 June 2010), https://www.webcitation.org/5qmWBEWcb?url=http://www.first14.com/bhopal-gas-tragedy-92-injuries-termed-minor-822.html

Fair, C. Christine, 'Pakistan's army is building an arsenal of "tiny" nuclear weapons – and it's going to backfire', *Quartz India* (21 December 2015), https://qz.com/579334/pakistans-army-is-building-an-arsenal-of-tiny-nuclear-weapons-and-its-going-to-backfire/

Federation of American Scientists, 'The Science of Military Strategy 2013', https://fas.org/nuke/guide/china/sms-2013.pdf

Fravel, M. Taylor & Medeiros, Evan S., 'China's Search for Assured Retaliation: The Evolution of Chinese Nuclear Strategy and Force Structure', *International Security*, vol. 35, No. 2 (2010)

Freedman, Lawrence, 'Nuclear Strategy', *Britannica*, https://www.britannica.com/topic/nuclear-strategy

Gady, Franz-Stefan, 'China, Russia Kick Off Anti-Ballistic Missile Defense Exercise', *The Diplomat* (12 December 2017), https://thediplomat.com/2017/12/china-russia-kick-off-anti-ballistic-missile-defense-exercise/

Gady, Franz-Stefan, 'India Successfully Tests New Ballistic Missile', *The Diplomat* (22 March 2016), http://thediplomat.com/2016/03/india-successfully-tests-new-ballistic-missile/

Ghoshal, Arkadev, 'India to Procure Anti-Chemical, Anti-Nuclear Suits from US for Rs 480 crore', *IB Times* (12 May 2017), http://www.ibtimes.co.in/india-procure-anti-chemical-anti-nuclear-suits-us-rs-480-crore-726410 (accessed 23 October 2017)

Godwin, Curt, 'Soyuz Rocket Successfully Delivers Eks-2 Early-Warning Satellite To Rare Orbit', *Spaceflight insider,* (25 May 2017), http://www.spaceflightinsider.com/organizations/roscosmos/soyuz-rocket-successfully-delivers-eks-2-early-warning-satellite-rare-orbit/

Graham, William, 'Atlas V Launches with SBIRS GEO-4', *Nasa Spaceflight* (19 January 2018), https://www.nasaspaceflight.com/2018/01/atlas-v-sbirs-geo-4-launch-cape-canaveral/

Gsponer, Andre & Hurni, Jean-Pierre, 'The Physical Principles of Thermonuclear Explosives, Inertial Confinement Fusion, and the Quest for Fourth Generation Nuclear Weapons', *Independent Scientific Research Institute Geneva-12, Switzerland* (20 January 2009)

Gupta, Shishir 'Govt okays induction of nuke-capable Shaurya missile amid Ladakh standoff', *Hindustan Times* (6 October 2020), https://www.hindustantimes.com/india-news/shaurya-missile-to-be-inducted-in-strategic-arsenal-agni-5-s-sea-version-by-2022/story-bS1100SkwoGLEXW5ANFQuO.html

Hawksley, Humphrey, 'India's Nuclear Muscle', *BBC World Service* (11 January 2003), http://news.bbc.co.uk/2/hi/south_asia/2646979.stm

http://pib.nic.in/archieve/lreleng/lyr2003/rjan2003/04012003/r040120033.html

https://defence.pk/pdf/threads/india-starts-building-nuclear-shelters.83066/

IDR Research Team, 'Operational Scenario Alpha', *Indian Defence Review* (July 1992)

Jaishankar, Dhruva, 'Decoding India's Nuclear Status', *Brookings* (3 April 2017), https://www.brookings.edu/opinions/decoding-indias-nuclear-status/

Jha, Saurav, 'DRDO's AAD Endo-atmospheric Ballistic Missile Interceptor Hits Bullseye', *Delhi Defence Review* (1 March 2017), http://www.delhidefencereview.com/2017/03/01/drdos-aad-endo-atmospheric-ballistic-missile-interceptor-hits-bullseye/

Jha, Saurav, 'Hit-to-Kill Successfully Demonstrated By DRDO's PDV Interceptor', *Delhi Defence Review* (25 February 2017), http://www.delhidefencereview.com/2017/02/25/hit-kill-successfully-demonstrated-drdos-pdv-interceptor/

Kampani, G 'New Delhi's Long Nuclear Journey', *International Security*, Vol. 38, No. 4 (Spring 2014)

Kampani, G., 'India's Evolving Civil-Military Institutions in an Operational Nuclear Context', *Carnegie Endowment for International Peace Regional Insight* (30 June 2016), https://carnegieendowment.org/2016/06/30/india-s-evolving-civil-military-institutions-in-operational-nuclear-context-pub-63910

Kanpur: Ammonia Gas Leak Causes Explosion in Cold Storage; Five Killed, Several Still Trapped', *First Post* (15 May 2017), http://www.firstpost.com/india/kanpur-ammonia-gas-leak-causes-explosion-in-cold-storage-five-killed-several-still-trapped-3335870.html

Kanwal, Gurmeet, 'Cold Start: India's Pro-active Offensive Operations Doctrine for War in the Plains', http://gurmeetkanwal.com/ArticleDetails.aspx?id=391

Kanwal, Gurmeet, 'Does India Need Tactical Nuclear Weapons?', *IDSA* (May 2003), http://www.idsa-india.org/an-may-03.html

Kanwal, Gurmeet, 'Does India Need Tactical Nuclear Weapons?', *Institute for Defence Studies and Analyses* (May 2003), https://www.idsa-india.org/an-may-03.html

Kanwal, Gurmeet, 'India's Cold Start Doctrine and Strategic Stability', *Institute for Defence Studies and Anayses* (1 June 2010), https://idsa.in/idsacomments/IndiasColdStartDoctrineandStrategicStability_gkanwal_010610

Kanwal, Gurmeet, 'India's Nuclear Force Structure 2025', *Carnegie Endowment for International Peace* (30 June 2016), http://carnegieendowment.org/2016/06/30/india-s-nuclear-force-structure-2025-pub-63988

Kanwal, Gurmeet, 'India's Nuclear Force Structure', *Institute for Defence Studies and Analyses* (September 2000), http://www.idsa-india.org/an-sept2-00.html

Kapila, Subhash, 'Indian Army Exercise Sudershan Shakti – Revalidation Of Cold Start War Doctrine', *South Asia Analysis Group* (10 December 2011), https://web.archive.org/web/20120104045741/http://southasiaanalysis.org/papers49/paper4809.html

Karber, Philip A., 'Strategic Implications of China's Underground Great Wall', *Georgetown University Asian Arms Control Project* (11 September 2011)

Karnad, Bharat, 'Going Thermonuclear: Why, With What Forces, at What Cost', *United Service Institution Journal*, vol. 17, No. 3 (July–September 1998), p.315

Kislyakov, Andrei, 'Russia to roll out new hypersonic missiles', *Russia Beyond* (18 December 2012), https://www.rbth.com/articles/2012/12/18/russia_to_roll_out_new_hypersonic_missiles_21107

Krishna, Ananatha M., 'K-15 SLBM is a Beast with Gen-Next Tech', *New Indian Express* (30 January 2013), http://www.newindianexpress.com/nation/2013/jan/30/k-15-slbm-is-a-beast-with-gen-next-tech-445756.html

Kristensen, Hans M. & Norris, Robert S., 'Chinese Nuclear Forces', *Bulletin of the Atomic Scientists*, vol. 72, No. 4 (2016)

Kristensen, Hans M. & Norris, Robert S., 'Indian Nuclear Forces, 2017', *Bulletin of the Atomic Scientists*, vol. 73, No. 4 (2017)

Kristensen, Hans M. & Norris, Robert S., 'Pakistani Nuclear Forces', *Bulletin of the Atomic Scientists*, vol. 72, No. 6 (2016)

Kulacki, Gregory, 'The Chinese Military Updates China's Nuclear Strategy', *Union of Concerned Scientists* (March 2015)

Kumar, Chethan, 'Six new Akash squadrons to give IAF missile muscle', *Times of India* (17 February 2015), https://timesofindia.indiatimes.com/india/Six-new-Akash-squadrons-to-give-IAF-missile-muscle/articleshow/46269673.cms

Kumar, R. Krishna, 'CBRN Disaster-Handling Centres to be Set Up in Four Metros', *The Hindu* (20 March 2016), http://www.thehindu.com/news/national/karnataka/cbrn-disasterhandling-centres-to-be-set-up-in-four-metros/article8376879.ece

Ladwig III, Walter C., 'A Cold Start for Hot Wars?', *International Security*, vol. 32, No. 3 (winter 2007/08), pp.158, http://www.mitpressjournals.org/doi/pdf/10.1162/isec.2008.32.3.158

Lin, Jeffrey & Singer, P.W., 'China Shot Down Another Missile in Space', *Popular Science* (13 February 2018), https://www.popsci.com/china-space-missile-test

Lodhi, Sardar F.S., 'Pakistan's Nuclear Doctrine', *Defence Journal* (April 1999), http://www.defencejournal.com/apr99/pak-nuclear-doctrine.htm

MacDonald, Bruce W. & Ferguson, Charles D., 'Chinese Strategic Missile Defense: Will It Happen, and What Would It Mean?', *Arms Control Today* (November 2015), https://www.armscontrol.org/ACT/2015_11/Features/Chinese-Strategic-Missile-Defense-Will-It-Happen-and-What-Would-It-Mean?utm_source=Arms+Control+Association+E-Updates&utm_campaign=114a736ce4-The_November_2015_ACT_Issue_Features&utm_medium=email&utm_term=0_0bf155a1a2-114a736ce4-79430281#notes

Magal, B.S., 'Guest Column: Checks and Balances: A Regular Review and Appraisal of the Nuclear Doctrine would Help to Fine Tune it As Per Strategic Requirements', *Force* (June 2014), http://forceindia.net/guest-column/guest-column-b-s-nagal/checks-and-balances/

Mallikarjun, Y. & Subramanian, T., 'Agni-V's Maiden Canister Trial a Roaring Success', *The Hindu* (31 January 2015), http://www.thehindu.com/news/national/maiden-canister-trial-of-agniv-a-roaring-success/article6841942.ece

Mallikarjun, Y. & Subramanian, T., 'Nirbhay Strays from Flight Path, Aborted', *The Hindu* (12 March 2013), http://www.thehindu.com/news/national/nirbhay-strays-from-flight-path-aborted/article4500527.ece

Mallikarjun, Y., 'Agni-II Missile Test-Fired Successfully', *The Hindu* (17 May 2010), http://www.thehindu.com/news/national/Agni-II-missile-test-fired-successfully/article16302660.ece (accessed 5 September 2017)

Mallikarjun, Y., 'Agni-III Test-Fired Successfully', *The Hindu* (21 September 2012), http://www.thehindu.com/news/national/agniiii-testfired-successfully/article3922230.ece

Mallikarjun, Y., 'Interceptor Missile Test Off Odisha Coast Fails', *The Hindu* (6 April 2015), http://www.thehindu.com/news/national/intercepter-missile-test-odisha-coast-wheeler-island drdo/article7073662.ece?ref=sliderNews (accessed 8 November 2017)

Mallikarjun, Y., 'Upgraded Interceptor Missile Successfully Hits Virtual Target', *The Hindu* (22 November 2015), http://www.thehindu.com/news/cities/Hyderabad/upgraded-interceptor-missile-successfully-engages-electronically-simulated-target-missile/article7905298.ece

Mallikarjun, Y., AMP & Subramanian, T., 'Agni-V Propels India into Elite ICBM Club', *The Hindu* (19 April 2012), http://www.thehindu.com/news/national/agniv-propels-india-into-elite-icbm-club/article3330921.ece?homepage=true

MacKenzie, D 'Making Waves', *New Scientist* (13 June 1998)

Menon, Air Marshal Narayan, 'Ballistic Missile Defence System for India', *Indian Defence Review,* vol. 27 (July–September 2012), http://www.indiandefencereview.com/spotlights/ballistic-missile-defence-system-for-india/

Menon, Raja, 'The Nuclear Doctrine: Yoking a Horse and Camel Together', *Times of India* (26 August 1999)

Mian, Zia, *et. al.*, *Plutonium Production in India and The US-India Nuclear Deal* (Washington DC: Nonproliferation Policy Education Center, 5 July 2005), http://www.npolicy.org/userfiles/image/Plutonium%20Production%20in%20India%20and%20the%20US-India%20Nuclear%20 Deal_pdf., pp.128–129

Mizokami, Kyle, 'Moscow's Defender: Russia Tests Modernized Ballistic Missile', *Popular Mechanics* (21 February 2018), https://www.popularmechanics.com/military/weapons/a18567468/moscows-defender-russia-tests-modernized-ballistic-missile/

Nagal, B.S., 'Guest Column: Nuclear No First Use Policy: A Time for Appraisal', *Force* (December 2014), http://forceindia.net/guest-column/guest-column-b-s-nagal/nuclear-no-first-use-policy/

Narang, Vipin, 'Plenary: Beyond the Nuclear Threshold: Causes and Consequences of First Use', *Carnegie International Nuclear Policy Conference*, Washington DC (20 March 2017), https://fbfy83yid9j1dqsev3zq0w8n-wpengine.netdna-ssl.com/wp-content/uploads/2013/08/Vipin-Narang-Remarks-Carnegie-Nukefest-2017.pdf

National Air and Space Intelligence Center (NASIC), 'Ballistic and Cruise Missile Threat' (2013), http://fas.org/programs/ssp/nukes/nuclearweapons/NASIC2013_050813.pdf

National Civil Defence College Website, 'Our Achievements', http://www.cddrm-ncdc.org/e39343/e41087/

National Disaster Management Authority, 'Capacity Building and Training'

National Disaster Management Authority, 'Evolution of NDMA', http://www.ndma.gov.in/en/about-ndma/evolution-of-ndma.html

National Disaster Management Authority, 'Fire Services', http://www.ndma.gov.in/en/fire-services.html.

National Disaster Management Authority, 'Functions and Responsibilities', http://www.ndma.gov.in/en/about-ndma/roles-responsibilities.html

National Disaster Management Authority, 'NDMA Roles and Responsibilities', http://www.ndma.gov.in/en/about-ndma/roles-responsibilities.html

National Disaster Management Authority, 'NDMA Vision', http://www.ndma.gov.in/en/about-ndma/vision.html

National Disaster Management Authority, 'Trauma Care', http://www.ndma.gov.in/en/capacity-building/trauma-care.html

National Disaster Management Plan May 2016 (New Delhi: NDMA, 2016), pp.26–27

NDRF Website, 'NDRF Training Regime', http://ndrf.gov.in/Media/inapnacms/TRAINING/taining%20regime.pdf

NDRF, 'About Us', http://ndrf.gov.in/ndrf

Nirbhay, India's Indigenous Cruise Missile, Fails Test Midway', *NDTV* (16 October 2015), http://www.ndtv.com/india-news/nirbhay-indias-indigenous-cruise-missile-fails-midway-1233086

Norris, Robert S. & Arkin, William M., 'Russian Nuclear Forces', *Bulletin of the Atomic Scientists*, vol. 56, No. 4 (2000), https://www.highbeam.com/doc/1G1-63794409.html

Nuclear Weapon Archive, 'Britain's Nuclear Weapons: British Nuclear Testing', http://nuclearweaponarchive.org/Uk/UKTesting.html

Nuclear Weapon Archive, 'Britain's Nuclear Weapons: History of the British Nuclear Arsenal', http://nuclearweaponarchive.org/Uk/UKArsenalDev.html

Nuclear Weapon Archive, 'Complete List of All U.S. Nuclear Weapons', http://nuclearweaponarchive.org/Usa/Weapons/Allbombs.html

Nuclear Weapon Archive, 'France's Nuclear Weapons: Development of the French Arsenal', http://nuclearweaponarchive.org/France/FranceArsenalDev.html

Pahwa, Pran, 'Minimum Deterrent: Defining the Concept', *Tribune* (11 February 1999)

Pahwa, Pran, *Organisation and Employment of Strategic Rocket Forces* (New Delhi: United Service Institution of India, 1999), pp.294–96

Panda, Ankit, 'India Inches Closer to Credible Nuclear Triad With K-4 SLBM Test', *The Diplomat* (13 May 2014), http://thediplomat.com/2014/05/india-inches-closer-to-credible-nuclear-triad-with-k-4-slbm-test/

Pandit, Rajat, 'After Agni-V Launch, DRDO's New Target is Anti-Satellite Weapons', *Times of India* (21 April 2012), https://timesofindia.indiatimes.com/india/After-Agni-

V-launch-DRDOs-new-target-is-anti-satellite-weapons/articleshow/12763074.cms

Pandit, Rajat. 'Ballistic missile Agni-IV test-fired as part of user trial', *Times of India* (9 November 2015), http://timesofindia.indiatimes.com/india/Ballistic-missile-Agni-IV-test-fired-as-part-of-user-trial/articleshow/49720522.cms

Pandit, Rajat, 'Agni-V with China in Range Tested; Next in Line is Agni-VI, with Multiple Warheads', *Times of India* (27 December 2016), http://timesofindia.indiatimes.com/india/agni-v-with-china-in-range-tested-next-in-line-is-agni-vi-with-multiple-warheads/articleshow/56191362.cms

Pandit, Rajat, 'India test-fires nuclear-capable ICBM Agni-V', *Times of India* (18 January 2018), https://timesofindia.indiatimes.com/india/india-test-fires-nuclear-capable-icbm-agni-v/articleshow/62550347.cms

Pandit, Rajat, 'Arihant's N-capable missile "ready to roll"', *Times of India* (25 January 2020), https://timesofindia.indiatimes.com/india/india-successfully-test-fires-k-4-submarine-launched-missile/articleshow/73589861.cms

Pant, Atul, 'EMP Weapons and the New Equation of War', *Institute for Defence Studies and Analyses* (13 October 2017), https://idsa.in/idsacomments/emp-weapons-new-equation-of-war_apant_131017

People's Republic of China Information Office of the State Council, 'China's National Defense in 2010' (31 March 2011)

People's Republic of China Information Office of the State Council, 'China's Military Strategy', *Xinhu* (26 May 2015), Strategic Guideline for Active Defense, http://eng.mod.gov.cn/DefenseNews/2015-05/26/content_4586748.htm

Pereira, Violet, 'N-capable Arihant submarine successfully test-fires unarmed missile', *Magalorean* (26 November 2015), http://www.mangalorean.com/n-capable-arihant-submarine-successfully-test-fires-unarmed-missile/

Peri, Dinakar, 'Now, India has a nuclear triad', *The Hindu* (18 October 2016), http://www.thehindu.com/news/national/Now-India-has-a-nuclear-triad/article16074127.ece

Philip, Snehesh Alex, 'India's ballistic missile shield ready, IAF & DRDO to seek govt nod to protect Delhi', *The Print* (8 January 2020), https://theprint.in/defence/indias-ballistic-missile-shield-ready-iaf-drdo-to-seek-govt-nod-to-protect-delhi/345853/.

Prakash, Arun, 'Strategic Policy Making and the Indian System', *Maritime Affairs*, vol. 5, No. 2 (Winter 2009), pp.22–31

Press Information Bureau, 'Pokhran-II Tests were Fully Successful; Given India Capability to build Nuclear Deterrence: Dr. Kakodkar and Dr. Chidambaram' (24 September 2009), http://pib.nic.in/newsite/PrintRelease.aspx?relid=52813

Press Information Bureau, 'Press Statement by Dr. Anil Kakodkar and Dr. R. Chidambaram on Pokhran-II tests' (24 September 2009), http://pib.nic.in/newsite/PrintRelease.aspx?relid=52814 (accessed 20 September 2017)

Press release No. PR16/2016-ISPR. Inter Services Public Relations' (19 January 2016), https://www.ispr.gov.pk/front/main.asp?o=t-press_release&date=2016/1/19

Press release No. PR248/2014-ISPR' (13 November 2014), https://www.ispr.gov.pk/front/main.asp?o=t-press_release&date=2014/11/13

Press release No. PR61/2015-ISPR. Inter Services Public Relations' (9 March 2015), https://www.ispr.gov.pk/front/main.asp?o=t-press_release&date=2015/3/9

Press release No. PR92/2014-ISPR' (15 April 2015), https://www.ispr.gov.pk/front/main.asp?o=t-press_release&date=2015/4/15

Press release PR122/2012-ISPR' (19 May 2012), https://www.ispr.gov.pk/front/main.asp?o=t-press_release&date=2012/5/19

PTI, 'Agni III could have 5,000km Range: Russian General', *The Hindu*, http://www.defencetalk.com/forums/missiles-wmds/indian-nuclear-missile-development-news-discussions-7241-11/

Pubby, Manu, 'India's Only Nuclear-Armed Submarine is Back in the Water after Contamination Fears', *The Print* (8 January 2018), https://theprint.in/2018/01/08/indias-only-nuclear-armed-submarine-ins-arihant-is-back-in-the-water/

Pubby, Manu, 'No "Cold Start" doctrine, India tells US', *Indian Express* (9 September 2010), http://archive.indianexpress.com/news/no--cold-start--doctrine-india-tells-us/679273/.#

Pubby, Manu, 'Setback for Indian missile programme: Two failures in a week, submarine version stuck', *The Print* (24 December 2017), https://theprint.in/2017/12/24/setback-for-indian-missile-programme-two-failures-in-a-week-submarine-version-stuck/ (accessed 10 January 2018)

Pubby, Manu, 'India's Rs 1.2 lakh crore nuclear submarine project closer to realization', *Economic Times* (21 February 2020), https://economictimes.indiatimes.com/news/defence/indias-rs-1-2-l-cr-n-submarine-project-closer-to-realisation/articleshow/74234776.cms

Raghuvanshi, V. 'Indian Excludes Foreign Vendors From Its Air-Defense Upgrade', *Defense News* (13 June 2016), https://www.defensenews.com/global/asia-pacific/2016/06/13/indian-excludes-foreign-vendors-from-its-air-defense-upgrade/

Rajagopalan, Rajeswari Pillai, 'India's Nuclear Security: Strengths and Gaps', *Observer Research Foundation* (14 June 2017), http://www.orfonline.org/research/india-nuclear-security-strengths-gaps/

Rasgotra, Maharajkrishna, 'Countering Nuclear Threats', in Brahma Chellaney (ed.), *Securing India's Future in the New Millennium* (New Delhi: Orient Longman, 1999)

Roberts, Brad, 'China and Ballistic Missile Defense: 1955 to 2002 and Beyond', *IDA Paper P-3826* (September 2003), p.21, n.80, http://www.dtic.mil/cgi-bin/GetTRDoc?AD=ADA418710 (citing Wan Yung-Kui, 'Can the Chinese Armed Forces Successfully Protect the Three-Gorges Dam?', *Hong Kong Tangai*, No. 31 [15 October 1993])

Robertson, Kalman A. & Carlson, John, 'The Three Overlapping Streams of India's Nuclear Programs', *Harvard Kennedy School, Belfer Centre for Science and International Affairs* (April 2016)

Roblin, Sebastien, 'America's Missile Defenses Against North Korea Have a Big Problem (They Only Work Half the Time)', *National Interest* (27 May 2017), http://nationalinterest.org/blog/the-buzz/americas-missile-defenses-against-north-korea-have-big-20872?page=show

Rout, Hemant Kumar, 'EXPRESS EXCLUSIVE: Maiden Test of Undersea K-4 Missile From Arihant Submarine', http://www.newindianexpress.com/nation/2016/apr/09/EXPRESS-EXCLUSIVE-Maiden-Test-of-Undersea-K-4-Missile-From-Arihant-Submarine-921990.html

Rout, Hemant Kumar, 'India successfully test fires Agni-V missile for a reduced range', *New Indian Express* (26 December 2016), http://www.newindianexpress.com/nation/2016/dec/26/india-successfully-test-fires-agni-v-missile-for-a-reduced-range-1553219.html

Rout, Hemant Kumar, 'India test fires new generation nuclear capable Agni-Prime missile off Odisha coast', *New Indian Express* (28 June 2021), https://www.newindianexpress.com/nation/2021/jun/28/india-test-fires-new-generation-nuclear-capable-agni-prime-missile-off-odisha-coast-2322550.html

Rout, Hemant Kumar, 'India to conduct first user trial of Agni-V missile', *New Indian Express* (13 September 2021), https://www.newindianexpress.com/states/odisha/2021/sep/13/india-to-conduct-first-user-trial-of-agni-v-missile-2357942.html

Saran, Shyam, 'Is India's Deterrent Credible?', *India Habitat Centre* (24 April 2013), http://www.armscontrolwonk.com/files/2013/05/Final-Is-Indias-Nuclear-Deterrent-Credible-rev1-2-1-3.pdf

Saran, Shyam, 'Is India's Nuclear Deterrent Credible?', Speech delivered at India Habitat Centre, New Delhi (24 April 2013)

Sawhney, P., 'If Pakistan and India Go to War', *The Asian Age* (12–13 November 1994)

Sawhney, Pravin, 'Decks Cleared For the Contract Signing of S-400 ADMS in December', *Force India* (17 October 2017), http://forceindia.net/decks-cleared-contract-signing-s-400-adms-december/

Sawhney, Pravin, 'How Inevitable is an Asian Missile Race?', *Jane's Intelligence Review* (January 2000), p.32

Sengupta, Prasun, 'Arudhra MPR Is EL/M-2084 MMR: Seeing Is Believing', *Trishul-Trident* (4 June 2011), http://trishul-trident.blogspot.com/2011/06/arudhra-mpr-is-elm-2084-mmr-seeing-is.html

Sengupta, Prasun, 'IAF's Multi-Phase IACCCS Being Enhanced', *Trishul-Trident* (22 January 2012), http://trishul-trident.blogspot.com/2012/01/iafs-multi-phase-iacccs-being-enhanced.html

Sengupta, Prasun, 'Poised for A Hattrick', *Force India* (October 2017), http://forceindia.net/cover-story/poised-for-a-hattrick/

Sharma, Suman, 'Army Mobilization Time: 48hrs', *DNA India* (16 May 2011), http://www.dnaindia.com/india/report-army-mobilisation-time-48-hours-1543679

Shrivastava, Kumar Sambhav, 'How Effective is India's Disaster Management Authority?', *Down to Earth* (20 June 2013), http://www.downtoearth.org.in/news/how-effective-is-india-s-disaster-management-authority--41415

Shukla, Ajai, 'Army Orders Surface to Air Missile, Making it the First Tri-Service Weapon', *Business Standard* (26 September 2017), http://www.business-standard.com/article/economy-policy/army-orders-surface-to-air-missile-making-it-the-first-tri-service-weapon-117092500988_1.html

Shukla, Ajai, 'India Launches 5,000-km Range Agni-5 Missile Successfully', *Business Standard* (20 April 2012), http://www.business-standard.com/article/economy-policy/india-launches-5-000-km-range-agni-5-missile-successfully-112042002020_1.html

Shukla, Ajai, 'New-Age Agni to boost Pak-Focused Nuclear Deterrent', *Business Standard* (17 December 2016), http://www.business-standard.com/article/economy-policy/new-age-agni-to-boost-pak-focused-nuclear-deterrent-116121601111_1.html

Shuster, Simon, 'Inside the Uranium Underworld: Dark Secrets, Dirty Bombs', *Time* (10 April 2017), http://time.com/4728293/uranium-underworld-dark-secrets-dirty-bombs/

Singh, Jasjit, 'A Nuclear Strategy for India', in Jasjit Singh (ed.), *Nuclear India* (New Delhi: Knowledge World, 1998)

Singh, Manvendra, 'Agni-II adds Firepower to N-Deterrence', *Indian Express* (12 April 1999)

Singh, Surendra, 'Military using 13 Satellites to Keep Eye on Foes', *Times of India* (26 June 2017), https://timesofindia.indiatimes.com/india/military-using-13-satellites-to-keep-eye-on-foes/articleshow/59314610.cms

Singh, Swaran, 'China's Nuclear Weapons and Doctrine', Raghuvanshi, Vivek, 'Indian Excludes Foreign Vendors From Its Air-Defense Upgrade', *Defense News* (13 June 2016), https://www.defensenews.com/global/asia-pacific/2016/06/13/indian-excludes-foreign-vendors-from-its-air-defense-upgrade/

Smithson, Amy E. & Levy, Leslie-Anne, 'Chapter 3 – Rethinking the Lessons of Tokyo', *Ataxia: The Chemical and Biological Terrorism Threat and the US Response* (Report), Henry L. Stimson Centre, pp.91, 95, 100. Report No. 35

Somasekhar, M. 'India's "Interceptor" May Make Ballistic Missile Shield Real', *The Hindu Business Line* (12 February 2017), http://www.thehindubusinessline.com/news/indias-interceptor-may-make-ballistic-missile-shield-real/article9537665.ece

Squassoni, Sharon, 'India's Nuclear Separation Plan: Issues and Views – CRS Report for Congress', *Congressional Research Service* (22 December 2006)

Sruthjith, K.K., 'Delhi Gradually Ushering in CWG Juggernaut', *Economic Times* (27 September 2010), https://economictimes.indiatimes.com/delhi-gradually-ushering-in-cwg-juggernaut/articleshow/6632785.cms

Stewart, Scott, 'Gauging the Threat of an Electromagnetic Pulse (EMP) Attack', *Stratfor Worldview* (9 September 2010), https://worldview.stratfor.com/article/gauging-threat-electromagnetic-pu

Subramanian T. & Mallikarjun, Y., 'Agni-II Soars in Success', *The Hindu* (30 September 2011), http://www.thehindu.com/news/national/agniii-soars-in-success/article2499781.ece

Subramanian, T. & Mallikarjun, Y., 'Agni-III test-fired successfully', *The Hindu* (8 May 2008), http://www.thehindu.com/todays-paper/Agni-III-test-fired-successfully/article15218480.ece

Subramanian, T. & Mallikarjun, Y., 'India successfully Test-Fires Shourya Missile', *The Hindu* (24 September 2011), http://www.thehindu.com/sci-tech/science/india-successfully-testfires-shourya-missile/article2482010.ece

Subramanian, T., 'Agni-IV Missile Successfully Test Fired', *The Hindu* (20 January 2014), http://www.thehindu.com/news/national/agniiv-missile-successfully-test-fired/article5596563.ece

Subramanian, T.S., 'Interceptor Missile Mission a "Failure"', *The Hindu* (23 May 2016), http://www.thehindu.com/news/national/Interceptor-missile-mission-a-%E2%80%98failure%E2%80%99/article14333898.ece

Subramanian, T.S., 'Nirbhay Likely to be Test-Fired in April', *The Hindu* (7 March 2012), http://www.thehindu.com/todays-paper/tp-national/nirbhay-likely-to-be-testfired-in-april/article2968219.ece

Subramanian, T.S., 'Nirbhay Missile Test "an utter failure"', *The Hindu* (21 December 2016), http://www.thehindu.com/news/national/Nirbhay-missile-test-%E2%80%9Can-utter-failure%E2%80%9D/article16915750.ece

Subramanian,T.S. 'Agni-IV missile successfully test fired', *The Hindu* (2 January 2017), http://www.thehindu.com/news/national/Agni-IV-missile-successfully-test-fired/article16977450.ece.

Subramanian, T.S., 'Smashing Hit', *Frontline* (22 December 2007–4 January 2008), http://www.frontline.in/static/html/fl2425/stories/20080104242512300.htm

Subramanyam, K., 'China and Nuclear Rationale', *Economic Times* (26 July 1997)

Subramanyam, Pillalmarri, 'Pakistan's Tactical Nuclear Weapons and India's Response', *Indian Defence Review Net Edition* (30 October 2016), http://www.indiandefencereview.com/news/pakistans-tactical-nuclear-weapons-and-indias-response/0/

Sundarji, General Krishnaswamy, 'Nuclear Deterrence Doctrine for India', *Trishul*, vol.5, No.2 (December 1992)

Sundarji, Krishnaswamy, 'Imperatives of Indian Minimum Deterrence', *Agni* (May 1996), p.21

Swami, Praveen, 'War and Games', *Frontline* (15 February 2002), http://www.frontline.in/static/html/fl1903/19030040.htm (accessed 23 October 2017)

Tellis, Ashley, 'Atoms for War? U.S.-Indian Civilian Nuclear Cooperation and India's Nuclear Arsenal', *Carnegie Endowment for International Peace* (2006), p.119. Zia Mian *et. al.*, 'Plutonium Production in India and the U.S.-India Nuclear Deal'

Thomas, M., 'An Analysis of the Threat Perception and Strategy for India', *Indian Defence Review* (January 1990)

Ting, Wang, 'Agni V and China/India Ballistic Missile Defense', presentation, Carnegie Endowment for International Peace, n.d., http://carnegieendowment.org/files/Wang_Ting%20 Presentation.pdf (presentation took place during week of 2 October 2012)

Tiwari, Mrigank, 'Unprecedented Security at Trishul Airbase', *Times of India* (3 January 2016), http://timesofindia.indiatimes.com/city/bareilly/Unprecedented-security-at-Trishul-airbase/articleshow/50430273.cms

Tiwary, Deeptiman, 'NDMA has failed to achieve objectives, task force says', https://timesofindia.indiatimes.com/india/NDMA-has-failed-to-achieve-objectives-task-force-says/articleshow/24119665.cms

Toucas, Boris, 'Ballistic Missile Defense: Proceed With Caution', *ArmsControl.org* (November 2017), https://www.armscontrol.org/act/2017-11/features/ballistic-missile-defense-proceed-caution

Trishul-Trident Blogspot (1 January 2018), http://trishul-trident.blogspot.in/2018/01/aad-endo-atmospheric-interceptor-headed.html

Tsuroka, Doug, 'China Pursuing Missile Defenses; Indian Nukes are Main Worry', *Asia Times* (19 January 2018), http://www.atimes.com/article/china-pursuing-missile-defenses-indian-nukes-main-worry/

Tur, Jatinder Kaur, 'India Developing E-bomb to Paralyze Networks', *Times of India* (29 August 2013), https://timesofindia.indiatimes.com/india/India-developing-E-bomb-to-paralyze-networks/articleshow/22127411.cms

Unnithan, Sandeep, 'From India Today magazine: A Peek into India's Top Secret and Costliest Defence Project, Nuclear Submarines', *India Today* (7 December 2017), http://indiatoday.intoday.in/story/india-ballistic-missile-submarine-k-6-submarine-launched-drdo/1/1104982.html

Unnithan, Sandeep, 'India has all the building blocks for an anti-satellite capability', *India Today* (27 April 2012), http://indiatoday.intoday.in/story/agni-v-drdo-chief-dr-vijay-kumar-saraswat-interview/1/186248.html

van Moyland, S. & Clark, R., 'The Paper Trail', *The Bulletin of the Atomic Scientists* (July/August 1998)

Varma, Roli & Varma, Daya R., 'The Bhopal Disaster of 1984', *Bulletin of Science, Technology and Society* vol.25, No.1 (2005)

Vasani, Harsh, 'India's Anti-Satellite Weapons', *The Diplomat* (14 June 2016), https://thediplomat.com/2016/06/indias-anti-satellite-weapons/

Visual Reverence, 'Orange Herald', http://visualreverence.tumblr.com/post/85541430823/orange-herald-was-a-fusion-boosted

Wadke, Rahul, 'Nuclear Plant to Have Eye in the Sky', *Hindu Business Line* (2 April 2017), http://www.thehindubusinessline.com/news/science/nuclear-plant-to-have-eye-in-the-sky/article9612475.ece

Weiner, Sarah, 'Recap: "China-India Nuclear Crossroads"', *Center for Strategic and International Studies* (9 October 2012), http://csis.org/blog/event-recap-china-india-nuclear-crossroads

Whitfield, Elizabeth, 'Fuzzy Math on Indian Nuclear Weapons', *Bulletin of the Atomic Scientists* (19 April 2017), http://thebulletin.org/fuzzy-math-indian-nuclear-weapons9343

Willman, David, '$40-Billion Missile Defense System Proves Unreliable', *La Times* (15 June 2014), http://www.latimes.com/nation/la-na-missile-defense-20140615-story.html

Zaidi, A. & Ehtisham, H., 'The Indian threat is real', *The Tribune.pk* (11 May 2017), https://tribune.com.pk/story/1406430/indian-threat-real/

Zhou, Laura, 'China and Russia Criticise THAAD Missile Defence System as Destabilising Region', *South China Morning Post* (9 July 2016), http://www.scmp.com/news/china/diplomacy-defence/article/1987103/china-and-russia-criticise-thaad-missile-defence-system

Books

Babbage, R. &. Gordon, S. (eds), *India's Strategic Future* (London: MacMillan, 1992)

Badri-Maharaj, Sanjay, *Indian Nuclear Strategy – Confronting the Potential Threat from China and Pakistan* (New Delhi: CLAWS, 2019)

Badri-Maharaj, Sanjay, *The Armageddon Factor: Nuclear Weapons in the India-Pakistan Context* (New Delhi: Lancer, 2000)

Bajpai, K., Chari, P., Cheema, S., Cohen S, & Ganguly, S. (eds), *Brasstacks and Beyond: Perception and Management of Crisis in South Asia* (New Delhi: Manohar, 1995)

Baranwal, J. ed., *SP's Military Yearbook 1992–93* (New Delhi: Guide Publications, 1993)

Burrows, W.E., & R. Windrem, *Critical Mass: The Dangerous Race for Superweapons in a Fragmenting World* (New York: Simon & Schuster, 1994)

Chari, P.R. *Indo-Pak Nuclear Standoff: The Role of the United States* (New Delhi: Manohar Publishers, 1995)

Chellaney, B. *Nuclear Proliferation: The US-Indian Conflict* (New Delhi: Orient Longman Ltd, 1993)

Chellaney, B. (ed.), *Securing India's Future in the New Millennium* (New Delhi: Orient Longman Limited, 1999)

Chengappa, Raj, *Weapons of Peace* (New Delhi: Harper Collins Publishers India, 2000)

Kanwal, Gurmeet, *Nuclear Defence: Shaping the Arsenal* (New Delhi: Knowledge World, 2001)

Karnad, Bharat (ed.), *Future Imperilled: India's Security in the 1990s and Beyond* (New Delhi: Viking Penguin India, 1994)

Karnad, Bharat, *India's Nuclear Policy* (Westport, Connecticut: Praeger Security International, 2008)

Karnad, Bharat, *Nuclear Weapons and Indian Security* (New Delhi: Macmillan India Ltd, 2002)

Karnad, Bharat, *Why India is not a Great Power (Yet)* (New Delhi: Oxford University Press, 2015)

Kavic, L.J., *India's Quest for Security* (London: University of London Press, 1967)

Menon, Shivshankar, *Choices: Inside the Making of India's Foreign Policy* (Penguin Random House India; Washington DC: Brookings, 2016)

Mirchandani, G. *India's Nuclear Dilemma* (New Delhi: Popular Book Services, 1968)

Nair, V.K., *Nuclear India* (New Delhi: Lancer International, 1992)

Palsokar, R.D, *Minimum Deterrent: India's Nuclear Answer to China* (Bombay: Thacker & Co. Ltd, 1969)

Perkovich, George, *India's Nuclear Bomb: The Impact on Global Proliferation* (London: University of California Press, 1999)

Sethi, Manpreet, *Nuclear Strategy: India's march towards Credible Deterrence* (New Delhi: KW Publishers, 2009)

Singh, Jasjit, 'A Nuclear Strategy for India', in Jasjit Singh, (ed.), *Nuclear India,* (New Delhi: Knowledge World in association with Institute for Defence Studies and Analyses, 1998)

Singh, S., *India and the Nuclear Bomb* (New Delhi: S. Chand & Co. Ltd, 1971)

Spector, L., *Nuclear Ambitions: The Spread of Nuclear Weapons 1989–90* (Boulder, Colorado: Westview Press, 1990)

Subramanyam, K. *Indian and the Nuclear Challenge* (New Delhi: Lancer, 1986)

NOTES

Introduction

1 L.J. Kavic, *India's Quest for Security* (London: University of London Press, 1967), p.19.
2 Kavic, p.41.
3 R.G.C. Thomas, 'The Growth of India's Military Powers', in R. Babbage & S. Gordon (eds), *India's Strategic Future* (London: MacMillan, 1992), p.39.
4 Kavic, p.37.
5 Kavic, p.96.
6 Thomas, 'Growth of Indian Military Power', p.43.
7 Thomas, 'Growth of Indian Military Power', pp.43–45.
8 W.E. Burrows & R. Windrem, *Critical Mass: The Dangerous Race for Superweapons in a Fragmenting World* (New York: Simon & Schuster, 1994), p.354.
9 B. Chellaney, *Nuclear Proliferation: The US-Indian Conflict* (New Delhi: Orient Longman Ltd, 1993), p.1.
10 Chellaney, *Nuclear Proliferation*, p.2.
11 Chellaney, *Nuclear Proliferation*, pp.1–8.
12 Chellaney, *Nuclear Proliferation*, p.4.
13 Chellaney, *Nuclear Proliferation*, p.5.
14 Leonard Spector, *Nuclear Ambitions: The Spread of Nuclear Weapons 1989–90* (Boulder, Colorado: Westview Press, 1990), p.86.
15 Spector, *Nuclear Ambitions*, p.87.
16 Chellaney, *Nuclear Proliferation*, pp.5–6.
17 Chellaney, *Nuclear Proliferation*, pp.5–6.
18 Chellaney, *Nuclear Proliferation*, p.7.
19 K. Subramanyam, *Indian and the Nuclear Challenge* (New Delhi: Lancer, 1986), p.222.
20 Spector, *Nuclear Ambitions*, p.64.
21 G. Mirchandani, *India's Nuclear Dilemma* (New Delhi: Popular Book Services, 1968), p.176.
22 R.D. Palsokar, *Minimum Deterrent: India's Nuclear Answer to China* (Bombay: Thacker & Co. Ltd, 1969), pp.57–58.
23 S. Singh, *India and the Nuclear Bomb* (New Delhi: S. Chand & Co. Ltd, 1971), p.135.
24 P.R. Chari, *Indo-Pak Nuclear Standoff: The Role of the United States* (New Delhi: Manohar Publishers, 1995), p.48.
25 'Smiling Buddha', *Nuclear Weapons Archive*, http://nuclearweaponarchive.org/India/IndiaSmiling.html.
26 D. Albright, 'The Shots Heard "Round the World"', *The Bulletin of the Atomic Scientists* (July/August 1998).
27 R. Chidambaran, 'Nuclear Dilemma', *India Today* (30 April 1994), p.50.
28 Chari, *Indo-Pak Nuclear Standoff*, pp.50–51.
29 'Shadow of an Indian H-Bomb', *Foreign Report* (13 December 1984), p.1.
30 Spector, *Nuclear Ambitions*, p.72.
31 Sanjay Badri-Maharaj, *The Armageddon Factor: Nuclear Weapons in the India-Pakistan Context* (New Delhi: Lancer, 2000), p.6.
32 See K. Bajpai, P.Chari, P. Cheema, S. Cohen & S. Ganguly (eds), *Brasstacks and Beyond: Perception and Management of Crisis in South Asia* (New Delhi: Manohar, 1995).
33 'The Long Pause 1974-1989', *Nuclear Weapon archive*, http://nuclearweaponarchive.org/India/IndiaPause.html.
34 R. Chengappa, *Weapons of Peace* (New Delhi: Harper Collins, 2000), p.305.
35 Chengappa, *Weapons of Peace*, p.304.
36 Chengappa, *Weapons of Peace*, p.304.
37 G. Perkovich, *India's Nuclear Bomb* (London: University of California Press, 1999), p.313.
38 Chengappa, *Weapons of Peace*, pp.382–85.
39 G. Kampani, 'New Delhi's Long Nuclear Journey', *International Security*, Vol. 38, No. 4 (Spring 2014), p.91.
40 'South Asian Nuclear-Weapon-Free Zone', *The Arms Control Reporter* (1993), p.454. B.176.

Chapter 1

1 'India's Nuclear Tests: News reports and "eyewitness" accounts', https://seismo.berkeley.edu/~rallen/research/nuke/India.May98/news.reports.html.
2 'NUCLEAR ANXIETY; Indian's Letter to Clinton On the Nuclear Testing'. *New York Times* (13 May 1998), https://www.nytimes.com/1998/05/13/world/nuclear-anxiety-indian-s-letter-to-clinton-on-the-nuclear-testing.html?auth=login-facebook.
3 'Operation Shakti', *Nuclear Weapons Archive.*
4 'Operation Shakti', *Nuclear Weapons Archive.*
5 'Operation Shakti', *Nuclear Weapons Archive.*
6 'India goes Thermonuclear', http://www.fas.org, p.2.
7 'India goes Thermonuclear', http://www.fas.org, p.2.
8 'India goes Thermonuclear', http://www.fas.org, p.2.
9 Wallace's report was discussed in the *LA Times*, 16 September 1998. An interesting point was made at the end of the article where an American nuclear weapons scientist indicated that even 15kt could have been a thermonuclear device and said that the seismologists had an inadequate knowledge of nuclear weapons design.
10 The reports about the 1974 test having a yield of only 2kt surfaced in an article in *The Sunday Observer* (30 August 1981) by Y. Aggarwal. While a few Western experts may cite this, there appears to be a consensus that the 1974 test had a yield of 8–12kt.
11 D. Albright, 'The Shots Heard "Round the World"', *The Bulletin of the Atomic Scientists* (July/August 1998).
12 'BARC scientists defend estimates of n-tests', *The Hindu* (19 September 1998).
13 'BARC scientists defend estimates of n-tests'.
14 'BARC scientists defend estimates of n-tests'.
15 D. MacKenzie, 'Making Waves', *New Scientist* (13 June 1998).
16 S. van Moyland & R. Clark, 'The Paper Trail', *The Bulletin of the Atomic Scientists* (July/August 1998).
17 S. van Moyland & R. Clark, 'The Paper Trail'.
18 F. Barnaby, 'Trials provide data for range of weapon yields', *Jane's Defence Weekly* 2 (7 May 1998), p.3.
19 F. Barnaby, 'Trials provide data for range of weapon yields', p.3.
20 F. Barnaby, 'Trials provide data for range of weapon yields', p.3.
21 'Operation Shakti', *Nuclear Weapons Archive.*
22 L. Spector, *Nuclear Ambitions* (Boulder: Westview, 1990), p.5; Albright & Hibbs, 'India's Silent Bomb', *The Bulletin of the Atomic Scientists* (September 1992), p.28, & 'Arms Control Register' (1993), p.176.
23 R. Chengappa, *Weapons of Peace*, p.437.
24 Chengappa, *Weapons of Peace*, p.327.
25 Chengappa, *Weapons of Peace*, pp.382–84.
26 Chengappa, *Weapons of Peace*, p.303.
27 G. Kampani, 'New Delhi's Long Nuclear Journey', *International Security*, Vol. 38, No. 4 (Spring 2014), p.91.

28 Kargil Review Committee, 'From Surprise to Reckoning', p.190. The committee reported that estimates of the 'number of cores/devices/ weapons in Pakistan's possession varied in different intelligence reports and assessments prepared for the government.
29 Kampani, 'New Delhi's Long Nuclear Journey', pp.91–92.
30 Kampani, 'New Delhi's Long Nuclear Journey', p.102.
31 Kampani, 'New Delhi's Long Nuclear Journey', p.103.
32 Kampani, 'New Delhi's Long Nuclear Journey', pp.103–04.
33 Kampani, 'New Delhi's Long Nuclear Journey', pp.103–04.
34 R. Chengappa, *Weapons of Peace*, pp.437–38.
35 'Pakistan was to deploy nukes against India during Kargil war', https://m.economictimes.com/news/defence/pakistan-was-to-deploy-nukes-against-india-during-kargil-war/articleshow/50019153.cms.

Chapter 2

1 For a more detailed discussion on the Indian nuclear doctrine, see Sanjay Badri-Maharaj, *Indian Nuclear Strategy: Confronting the Potential Threat from both China and Pakistan* (New Delhi: Centre for Land Warfare Studies, 2019), pp.240–278.
2 'Draft Indian Nuclear Doctrine', *Arms Control.org*, https://www.armscontrol.org/act/1999_07-08/ffja99.
3 'CABINET COMMITTEE ON SECURITY REVIEWS PROGRESS IN OPERATIONALIZING INDIA'S NUCLEAR DOCTRINE', http://pib.nic.in/archieve/lreleng/lyr2003/rjan2003/04012003/r040120033.html.
4 A. Ahmed, 'The Need For Clarity In India's Nuclear Doctrine', *Institute for Defence Studies and Analyses*, (11 November 2008), https://idsa.in/idsastrategiccomments/TheNeedForClarityInIndiaSNuclearDoctrine_AAhmed_111108.
5 A. Ahmed, 'The Need For Clarity In India's Nuclear Doctrine'.
6 S. Chandra, 'Revisiting India's Nuclear Doctrine: Is it necessary?', *Institute for Defence Studies and Analyses* (30 April 2014), https://idsa.in/issuebrief/RevisitingIndiasNuclearDoctrine_schandra_300414.
7 S. Chandra, 'Revisiting India's Nuclear Doctrine: Is it necessary?'.
8 Badri-Maharaj, *Indian Nuclear Strategy*, pp.267–274.
9 Vipin Narang, 'Plenary: Beyond the Nuclear Threshold: Causes and Consequences of First Use', Carnegie International Nuclear Policy Conference, Washington DC (20 March 2017), https://fbfy83yid9j1dqsev3zq0w8n-wpengine.netdna-ssl.com/wp-content/uploads/2013/08/Vipin-Narang-Remarks-Carnegie-Nukefest-2017.pdf.
10 Shivshankar Menon, *Choices: Inside the Making of India's Foreign Policy* (Washington DC: Brookings, 2016), p.117.
11 Menon, *Choices*, pp.105–123.
12 Menon, *Choices*, p.117.
13 Badri-Maharaj, *Indian Nuclear Strategy*, pp.257–267.
14 S. Saran, *Is India's Deterrent Credible?*, *India Habitat Centre* (24 April 2013), http://www.armscontrolwonk.com/files/2013/05/Final-Is-Indias-Nuclear-Deterrent-Credible-rev1-2-1-3.pdf.
15 S. Saran, *Is India's Deterrent Credible?*.
16 'India building nuclear-proof bunkers for top leaders', *Silicon India* (23 September 2003), http://www.siliconindia.com/shownews/India_building_nuclearproof_bunkers_for_top_leaders-nid-20887-cid-Top.html.
17 https://defence.pk/pdf/threads/india-starts-building-nuclear-shelters.83066/.
18 Karnad, *India's Nuclear Policy*, p.102.
19 S. Chandra, 'Revisiting India's Nuclear Doctrine: Is it necessary?'.
20 Gaurav Kampani, 'India's Evolving Civil-Military Institutions in an Operational Nuclear Context', *Carnegie Endowment for International Peace Regional Insight* (30 June 2016), https://carnegieendowment.org/2016/06/30/india-s-evolving-civil-military-institutions-in-operational-nuclear-context-pub-63910.
21 Kampani, 'India's Evolving Civil-Military Institutions in an Operational Nuclear Context'.
22 S. Chandra, 'Revisiting India's Nuclear Doctrine: Is it necessary?'.
23 https://defence.pk/pdf/threads/india-starts-building-nuclear-shelters.83066/.
24 https://defence.pk/pdf/threads/india-starts-building-nuclear-shelters.83066/.
25 Dhruv Jaishankar, 'Decoding India's Nuclear Status', *Brookings* (3 April 2017), https://www.brookings.edu/opinions/decoding-indias-nuclear-status/.
26 S. Chandra, 'Revisiting India's Nuclear Doctrine: Is it necessary?'.
27 A. Ahmed, 'The Need For Clarity In India's Nuclear Doctrine'.
28 Badri-Maharaj, *Indian Nuclear Strategy*, pp.209–240.
29 'Government approves appointment of 3 experts as NDMA members', *Press Trust of India* (21 December 2014), http://www.dnaindia.com/india/report-government-approves-appointment-of-3-experts-as-ndma-members-2045959 (accessed 10 November 2017).
30 'Government to restructure and revamp NDMA', *GK Today* (30 September 2014), https://currentaffairs.gktoday.in/government-restructure-revamp-ndma-09201414966.html (accessed 10 November 2017).
31 'National Disaster Management Plan Part 1: Draft', p.61.

Chapter 3

1 Badri-Maharaj, *Indian Nuclear Strategy*, pp.31–97.
2 Badri-Maharaj, *Indian Nuclear Strategy*, pp.31–97.
3 Sharon Squassoni, 'India's Nuclear Separation Plan: Issues and Views – CRS Report for Congress', *Congressional Research Service* (22 December 2006), pp.20–21.
4 Sanjay Badri-Maharaj, *Indian Nuclear Strategy: Confronting the Potential Threat from both China and Pakistan* (New Delhi: Centre for Land Warfare Studies, 2019), pp.79–84.
5 Kalman A. Robertson & John Carlson, *The Three Overlapping Streams of India's Nuclear Programs* (Harvard Kennedy School: Belfer Centre for Science and International Affairs, April 2016), pp.13–19.
6 'India has fissile material for 2,000 warheads: Pak media', *Times of India* (10 September 2015), http://timesofindia.indiatimes.com/world/pakistan/India-has-fissile-material-for-2000-warheads-Pak-media/articleshow/48895568.cms.
7 E. Whitfield, 'Fuzzy Math on Indian Nuclear Weapons', *The Bulletin of the Atomic Scientists* (19 April 2017), http://thebulletin.org/fuzzy-math-indian-nuclear-weapons9343.
8 Whitfield, 'Fuzzy Math on Indian Nuclear Weapons'.
9 'Third Pakistani nuclear reactor operational: US magazine', *The Nation* (7 July 2014). http://nation.com.pk/islamabad/07-Jul-2014/third-pakistani-nuclear-reactor-operational-us-magazine.
10 Whitfield, 'Fuzzy Math on Indian Nuclear Weapons'.
11 Whitfield, 'Fuzzy Math on Indian Nuclear Weapons'.
12 Bharat Karnad, *India's Nuclear Policy* (Westport, Connecticut: Praeger Security International, 2008), p.84. In an interview with Professor Karnad, Dr Iyengar refutes Ashley Tellis's assertions that India could not reprocess 55–60kg of plutonium per year and said that figure had already been achieved.
13 E. Whitfield, 'Fuzzy Math on Indian Nuclear Weapons'.
14 While undoubtedly some of India's HEU will be used for these reactors, given that India's SSBN and SSN projects are moving at a ponderous pace, this assertion might be an overstatement.
15 Bharat Karnad, *India's Nuclear Policy*, p.73.
16 Andre Gsponer & Jean-Pierre Hurni, *The physical principles of thermonuclear explosives, inertial confinement fusion, and the quest for four-generation nuclear weapons* (Geneva, Switzerland: Independent Scientific Research Institute, 20 January 2009), p.17.
17 M. Ahmed, *India's Nuclear Exceptionalism, Fissile Materials, Fuel Cycles, and Safeguards* (Harvard Kennedy School: Belfer Centre for Science and International Affairs, Discussion Paper, May 2017), pp.51–53.
18 Ahmed's estimates consider 4kg for WG Pu, 8kg for RG Pu and 20kg for WG HEU. Calculating the weapons potential for India's fissile materials presumes that all of its unsafeguarded stockpile has military potential; India has yet to commission its first PFBR, which keeps the proliferation potential of all of India's separated and unseparated weapon-usable plutonium stockpile intact; India has declared its unsafeguarded RG Pu stockpile as 'strategic'; India's reprocessing capacity is steadily increasing in size and efficiency; and India's unsafeguarded HEU stockpile can easily be scaled up to weapons-grade levels, notwithstanding that its plans for a fleet of five nuclear submarines is far from complete.
19 A. Tellis, 'Atoms for War? US-Indian Civilian Nuclear Cooperation and India's Nuclear Arsenal' (Carnegie Endowment for International Peace, 2006), pp.19–27.
20 Ahmed, *India's Nuclear Exceptionalism*, n. 126.
21 G. Perkovich, *India's Nuclear Bomb*, pp.429–430.
22 Bharat Karnad, *Nuclear Weapons and Indian Security* (New Delhi: Macmillan India Ltd, 2002), pp.20–23.
23 Zia Mian, *et. al.*, *Plutonium Production in India and The US-India Nuclear Deal* (Washington DC: Nonproliferation Policy Education Center, 5 July 2005), http://www.npolicy.org/userfiles/ image/Plutonium%20Production%20in%20India%20and%20the%20US-India%20Nuclear%20 Deal_pdf., pp.128–129.
24 Bharat Karnad, *India's Nuclear Policy*, pp.92–93.

25 Karnad, *India's Nuclear Policy*, p.84.
26 Badri-Maharaj, *Indian Nuclear Strategy*, pp.56–65.
27 Bharat Karnad, *Nuclear Weapons and Indian Security* (New Delhi: Macmillan India Ltd, 2002), p.320.
28 Karnad, *Nuclear Weapons*, p.376.
29 'India capable of building nuke deterrence up to 200 kilotons: Kakodkar', *Deccan Herald* (24 September 2009), http://www.deccanherald.com/content/27047/india-capable-building-nuke-.
30 G. Perkovich, *India's Nuclear Bomb*, p.242.
31 Bharat Karnad, *India's Nuclear Policy*, pp.80–95.
32 A. Prakash, 'Strategic Policy Making and the Indian System', *Maritime Affairs*, Vol. 5 No. 2 (Winter 2009), pp.22–31; at p.22 Admiral Prakash responded to a query by the author by stating that weapons in the 200–500kt yield were part of the Indian arsenal and assertions that India's reliable arsenal was limited to 20–25kt weapons was "inaccurate by an order of magnitude".
33 A. Shukla, 'India launches 5,000-km range Agni-5 missile successfully', *Business Standard* (20 April 2012), http://www.business-standard.com/article/economy-policy/india-launches-5-000-km-range-agni-5-missile-successfully-112042002020_1.html.
34 'Press Statement by Dr. Anil Kakodkar and Dr. R. Chidambaram on Pokhran-II tests', *Press Information Bureau* (24 September 2009), http://pib.nic.in/newsite/PrintRelease.aspx?relid=52814 (accessed 20 September 2017); see also 'Pokhran-II tests were fully successful; given India capability to build nuclear deterrence: Dr. Kakodkar and Dr. Chidambaram, *Press Information Bureau* (24 September 2009), http://pib.nic.in/newsite/PrintRelease.aspx?relid=52813.
35 Bharat Karnad, *India's Nuclear Policy*, p.68.
36 Karnad, *India's Nuclear Policy*, p.78.
37 Karnad, *Nuclear Weapons*, pp.20–23.
38 Bharat Karnad, *Why India is not a Great Power (Yet)* (New Delhi: Oxford University Press, 2015), p.370.
39 'Use plural, India has thermonuclear bombs: Kakodkar', *CNN-IBN Interview Transcript* (13 December 2009).
40 Karnad, *India's Nuclear Policy*, p.82.
41 Karnad, *India's Nuclear Policy*, p.82.
42 F. Barnaby, 'Trials provide data for range of weapon yields', p.3.
43 Badri-Maharaj, *Indian Nuclear Strategy*, pp.56–65.
44 H. Hawksley, 'India's nuclear muscle', *BBC World Service* (11 January 2003), http://news.bbc.co.uk/2/hi/south_asia/2646979.stm.
45 Karnad, *India's Nuclear Policy*, p.99.
46 S. Saran, 'Is India's Nuclear Deterrent Credible?'.
47 Hans M. Kristensen & Robert S. Norris, 'Indian nuclear forces, 2017', *The Bulletin of the Atomic Scientists*, 73:4 (2017), pp.205–09.
48 Kristensen & Norris, p.206.
49 Karnad, *India's Nuclear Policy*, pp.89–93.
50 G. Kanwal, *Nuclear Defence: Shaping the Arsenal* (New Delhi: Knowledge World, 2001), pp.133–137.
51 G. Kanwal, 'India's Nuclear Force Structure 2025', *Carnegie Endowment for International Peace* (30 June 2016), http://carnegieendowment.org/2016/06/30/india-s-nuclear-force-structure-2025-pub-63988.
52 K. Subramanyam, 'Nuclear Force Design and Minimum Deterrence Strategy for India', in Bharat Karnad (ed.), *Future Imperilled: India's Security in the 1990s and Beyond* (New Delhi: Viking Penguin India, 1994), pp.189–193.
53 K. Subramanyam, 'China and Nuclear Rationale', *Economic Times* (26 July 1997).
54 Subramanyam, 'China and Nuclear Rationale'.
55 Maharaja Krishna Rasgotra, 'Countering Nuclear Threats', in Brahma Chellaney (ed.), *Securing India's Future in the New Millennium* (New Delhi: Orient Longman, 1999), pp.238–239.
56 K. Sundarji, 'Imperatives of Indian Minimum Deterrence', *Agni* (May 1996), p.21.
57 V.K. Nair, *Nuclear India* (New Delhi: Lancer International, 1992), pp.170–182.
58 Raja Menon, 'The Nuclear Doctrine: Yoking a Horse and Camel Together', *Times of India* (26 August 1999).
59 Brigadier Kanwal's account of Subramanian's assessment was done by interview.
60 Bharat Karnad, 'Going Thermonuclear: Why, With What Forces, at What Cost', *United Service Institution Journal* 17, No. 3 (July–September 1998), p.315.
61 Karnad, 'Going Thermonuclear'.
62 Pran Pahwa, *Organisation and Employment of Strategic Rocket Forces* (New Delhi: United Service Institution of India, 1999), pp.294–296.
63 Pran Pahwa, 'Minimum Deterrent: Defining the Concept', *Tribune* (11 February 1999). Kanwal, 'India's Nuclear Force Structure', *Institute for Defence Studies and Analyses* (September 2000), http://www.idsa-india.org/an-sept2-00.html.
64 G. Kanwal, *Nuclear Defence: Shaping the Arsenal* (New Delhi: Knowledge World, 2001), p.96.
65 Kanwal, *Nuclear Defence*, p.96.
66 G. Kanwal, 'Does India Need Tactical Nuclear Weapons?', *IDSA* (May 2003), http://www.idsa-india.org/an-may-03.html (accessed 19 September 2017).
67 S. Badri-Maharaj, 'The Importance of Passive and Active CBRN Defensive Measures', *Institute for Defence Studies and Analyses: Issue Brief* (17 October 2016), https://idsa.in/issuebrief/importance-of-passive-and-active-cbrn-defensive-measures_sbmaharaj_171016 (accessed 23 October 2017).
68 S.K. Abbas, 'Brahmos anti-ship missile to be produced in 2003', *Rediff.com* (1 June 2002), http://www.rediff.com/news/2002/jun/01war11.htm.
69 'Indian army may soon get bio-chem suits', *Rediff.com* (11 May 2011), http://www.rediff.com/news/report/indian-army-may-soon-get-bio-chem-suits/20110511.htm.
70 A. Ghoshal, 'India to procure anti-chemical, anti-nuclear suits from US for Rs 480 crore', *IB Times* (12 May 2017), http://www.ibtimes.co.in/india-procure-anti-chemical-anti-nuclear-suits-us-rs-480-crore-726410.
71 G. Perkovich, *India's Nuclear Bomb*, p.242.
72 B. Karnad, *Nuclear Weapons*, p.320.
73 R. Chengappa, *Weapons of Peace*, p.327.
74 Chengappa, *Weapons of Peace*, pp.382–384.
75 S. Aroor, '10 reasons why the Indian Rafale is evolution itself', *Daily O* (4 July 2017), http://www.dailyo.in/variety/rafale-aircraft-brahmos-nuclear-defence/story/1/18157.html.
76 M. Tiwari, 'Unprecedented security at Trishul airbase', *Times of India* (3 January 2016), http://timesofindia.indiatimes.com/city/bareilly/Unprecedented-security-at-Trishul-airbase/articleshow/50430273.cms.
77 'India Plans Next Generation Bombproof Shelter for 108 Fighter Jets' *Sputnik News* (3 July 2017), https://sputniknews.com/military/201707031055187413-india-shelter-jets/ (accessed 4 September 2017).
78 Bharat Karnad, *India's Nuclear Policy*, p.82.
79 S. Aroor, '"True" BrahMos Unleashed Today, Next 900-km Weapon' *Livefist Defence*, https://www.livefistdefence.com/2017/03/true-brahmos-unleashed-today-next-1000-km-weapon.html.
80 T.S. Subramanian, 'Nirbhay likely to be test-fired in April', *The Hindu* (7 March 2012), http://www.thehindu.com/todays-paper/tp-national/nirbhay-likely-to-be-testfired-in-april/article2968219.ece.
81 Y. Mallikarjun & T. Subramanian, 'Nirbhay strays from flight path, aborted', *The Hindu* (12 March 2013), http://www.thehindu.com/news/national/nirbhay-strays-from-flight-path-aborted/article4500527.ece.
82 'India test-fires nuclear-capable Nirbhay cruise missile', *Times of India* (17 October 2014), http://timesofindia.indiatimes.com/india/India-test-fires-nuclear-capable-Nirbhay-cruise- missile/articleshow/44845526.cms.
83 'Nirbhay, India's Indigenous Cruise Missile, Fails Test Midway', *NDTV* (16 October 2015), http://www.ndtv.com/india-news/nirbhay-indias-indigenous-cruise-missile-fails-midway-1233086.
84 T.S. Subramanian, 'Nirbhay missile test "an utter failure"', *The Hindu* (21 December 2016), http://www.thehindu.com/news/national/Nirbhay-missile-test-%E2%80%9Can-utter-failure%E2%80%9D/article16915750.ece.
85 'Sub-sonic cruise missile "Nirbhay" successfully test-fired', *The Hindu* (16 April 2019), https://www.thehindu.com/news/national/other-states/sub-sonic-cruise-missile-nirbhay-successfully-test-fired/article26849032.ece.
86 Shiv Aroor, 'India deploys long-range missile Nirbhay to counter Chinese threat at LAC', *India Today* (28 September 2020), https://www.indiatoday.in/india/story/india-deploys-long-range-missile-nirbhay-to-counter-chinese-threat-at-lac-1726255-2020-09-28.
87 Badri-Maharaj, *Indian Nuclear Strategy*, pp.36–53.
88 Karnad, *India's Nuclear Policy*, p.78.
89 Chengappa, *Weapons of Peace*, p.437.
90 Karnad, *India's Nuclear Policy*, p.83 n.114.
91 R. Chengappa, 'The Missile Man', *India Today* (15 April 1994), pp.40–42.
92 M. Singh, 'Agni-II adds Firepower to N-Deterrence', *Indian Express* (12 April 1999).

93 Y. Mallikarjun, 'Agni-II missile test-fired successfully', *The Hindu* (17 May 2010), http://www.thehindu.com/news/national/Agni-II-missile-test-fired-successfully/article16302660.ece.

94 T. Subramaniam & Y. Mallikarjun, 'Agni-II soars in success', *The Hindu* (30 September 2011), http://www.thehindu.com/news/national/agniii-soars-in-success/article2499781.ece.

95 'Odisha: Nuclear capable Agni-II missile successfully test fired', *IBN Live* (7 April 2013), https://archive.is/20130628175643/http://www.bharat-rakshak.com/NEWS/newsrf.php.

96 'India successfully test-fires nuclear capable Agni-II missile off Odisha coast', *News 18* (9 November 2014), http://www.news18.com/news/india/india-successfully-test-fires-nuclear-capable-agni-ii-missile-off-odisha-coast-724952.html.

97 'India test fires medium range nuclear capable Agni-II missile', *Economic Times* (20 February 2018), https://economictimes.indiatimes.com/news/defence/india-test-fires-agni-ii-missile-off-odisha-coast/articleshow/62993975.cms?utm_source=contentofinterest&utm_medium=text&utm_campaign=cppst.

98 'India conducts first night trial of Agni-II missile', *India Today* (16 November 2019), https://www.indiatoday.in/india/story/india-conducts-first-night-trial-of-agni-ii-missile-1619704-2019-11-16.

99 'DRDO told to test Upgraded Agni in August', *Deccan Chronicle* (15 July 1998).

100 'India Successfully Test-fires N-capable Agni-I Ballistic Missile', *News 18* (22 November 2016), http://www.news18.com/news/india/india-successfully-tests-nuclear-capable-agni-i-ballistic-missile-1314397.html.

101 Hemant Kumar Rout, 'India test fires new generation nuclear capable Agni-Prime missile off Odisha coast', *New Indian Express* (28 June 2021), https://www.newindianexpress.com/nation/2021/jun/28/india-test-fires-new-generation-nuclear-capable-agni-prime-missile-off-odisha-coast-2322550.html.

102 T. Subramanian & Y. Mallikarjun, 'Agni-III test-fired successfully', *The Hindu* (8 May 2008), http://www.thehindu.com/todays-paper/Agni-III-test-fired-successfully/article15218480.ece.

103 Y. Mallikarjun, 'Agni-III test-fired successfully', *The Hindu* (21 September 2012), http://www.thehindu.com/news/national/agniiii-testfired-successfully/article3922230.ece.

104 'Agni-III successfully test fired from Odisha coast', *Economic Times* (16 April 2015), http://economictimes.indiatimes.com/news/defence/agni-iii-successfully-test-fired-from-odisha-coast/articleshow/46941664.cms.

105 'India test-fires nuclear-capable Agni III ballistic missile', *Economic Times* (27 April 2017), http://economictimes.indiatimes.com/news/defence/india-test-fires-nuclear-capable-agni-iii-ballistic-missile/articleshow/58396268.cms.

106 'Agni III could have 5,000 km range: Russian General', *PTI* article carried in *The Hindu*, http://www.defencetalk.com/forums/missiles-wmds/indian-nuclear-missile-development-news-discussions-7241-11/.

107 'India to test fire Agni-V by year-end', *The Hindu* (3 June 2011), http://www.thehindu.com/sci-tech/science/India-to-test-fire-Agni-V-by-year-end/article13821309.ece.

108 'Long range strategic missile Agni-IV test-fired', *The Hindu* (19 September 2012), http://www.thehindu.com/sci-tech/science/long-range-strategic-missile-agniiv-testfired/article3914340.ece.

109 T. Subramanian, 'Agni-IV missile successfully test fired', *The Hindu* (20 January 2014), http://www.thehindu.com/news/national/agniiv-missile-successfully-test-fired/article5596563.ece.

110 'India successfully tests nuclear capable Agni IV missile', *The Hindu* (3 December 2014), http://www.thehindu.com/todays-paper/tp-in-school/india-successfully-tests-nuclear-capable-agni-iv-missile/article6656285.ece.

111 R. Pandit, 'Ballistic missile Agni-IV test-fired as part of user trial', *Times of India* (9 November 2015), http://timesofindia.indiatimes.com/india/Ballistic-missile-Agni-IV-test-fired-as-part-of-user-trial/articleshow/49720522.cms.

112 T. Subramanian, 'Agni-IV missile successfully test fired', *The Hindu* (2 January 2017), http://www.thehindu.com/news/national/Agni-IV-missile-successfully-test-fired/article16977450.ece.

113 'Nuclear Capable Agni-IV missile successfully test fired', *Economic Times* (23 December 2018), https://economictimes.indiatimes.com/news/defence/nuclear-capable-agni-iv-missile-successfully-test-fired/articleshow/67215177.cms?from=mdr.

114 Y. Mallikarjun, AMP & T. Subramanian, 'Agni-V propels India into elite ICBM club', *The Hindu* (19 April 2012), http://www.thehindu.com/news/national/agniv-propels-india-into-elite-icbm-club/article3330921.ece?homepage=true.

115 'India test-fires Agni V with range as far as China', *Hindustan Times* (16 September 2013), http://www.hindustantimes.com/india/india-test-fires-agni-v-with-range-as-far-as-china/story-28IHgfrhxGgUt9XLNPRiwN.html.

116 Y. Mallikarjun & T. Subramanian, 'Agni-V's maiden canister trial a roaring success', *The Hindu* (31 January 2015), http://www.thehindu.com/news/national/maiden-canister-trial-of-agniv-a-roaring-success/article6841942.ece.

117 H.K. Rout, 'India successfully test fires Agni-V missile for a reduced range', *New Indian Express* (26 December 2016), http://www.newindianexpress.com/nation/2016/dec/26/india-successfully-test-fires-agni-v-missile-for-a-reduced-range-1553219.html.

118 'Agni-V can reach targets 8,000 km away: Chinese expert', *The Hindu* (21 April 2012), http://www.thehindu.com/todays-paper/tp-in-school/agniv-can-reach-targets-8000-km-away-chinese-expert/article3337202.ece.

119 'India test fires nuclear capable Agni-5 missile, 2nd test in six months', *Economic Times* (10 December 2018), https://economictimes.indiatimes.com/news/defence/india-test-fires-nuclear-capable-agni-5-missile-2nd-test-in-six-months/articleshow/67023684.cms.

120 Hemant Kumar Rout, 'India to conduct first user trial of Agni-V missile', *New Indian Express* (13 September 2021), https://www.newindianexpress.com/states/odisha/2021/sep/13/india-to-conduct-first-user-trial-of-agni-v-missile-2357942.html.

121 R. Pandit, 'Agni-V with China in range tested; next in line is Agni-VI, with multiple warheads', *Times of India* (27 December 2016), http://timesofindia.indiatimes.com/india/agni-v-with-china-in-range-tested-next-in-line-is-agni-vi-with-multiple-warheads/articleshow/56191362.cms.

122 A. Shukla, "New-age Agni to boost Pak-focused nuclear deterrent', *Business Standard* (17 December 2016), http://www.business-standard.com/article/economy-policy/new-age-agni-to-boost-pak-focused-nuclear-deterrent-116121601111_1.html.

123 T. Subramanian & Y. Mallikarjun, 'India successfully test-fires Shourya missile', *The Hindu* (24 September 2011), http://www.thehindu.com/sci-tech/science/india-successfully-testfires-shourya-missile/article2482010.ece.

124 'Prithvi missiles to be replaced by more-capable Prahar: DRDO', *The Hindu Business Line* (30 June 2013), http://www.thehindubusinessline.com/news/prithvi-missiles-to-be-replaced-by-morecapable-prahar-drdo/article4866081.ece.

125 Shishir Gupta 'Govt okays induction of nuke-capable Shaurya missile amid Ladakh standoff', *Hindustan Times* (6 October 2020), https://www.hindustantimes.com/india-news/shaurya-missile-to-be-inducted-in-strategic-arsenal-agni-5-s-sea-version-by-2022/story-bS1100SkwoGLEXW5ANFQuO.html.

126 'India to join ICBM club soon – Interview with Dr. S. Christopher', *NDTV* (11 July 2015), http://www.ndtv.com/video/news/news/india-to-join-icbm-club-soon-374683.

127 The only other indication of production levels came in an interview Dr Saraswat gave in 2012 to *India Today* where he said, 'All I can tell you is that we will produce more than just 1 or 2 missiles a year.' S. Unnithan, 'India has all the building blocks for an anti-satellite capability' *India Today* (27 April 2012), http://indiatoday.intoday.in/story/agni-v-drdo-chief-dr-vijay-kumar-saraswat-interview/1/186248.html.

128 'Ballistic and Cruise Missile Threat 2017', *Defense Intelligence Ballistic Missile Analysis Committee*, http://www.nasic.af.mil/Portals/19/images/Fact%20Sheet%20Images/2017%20Ballistic%20and%20Cruise%20Missile%20Threat_Final_small.pdf?ver=2017-07-21-083234-343, p.25.

129 Karnad, *India's Nuclear Policy*, p.101.

130 Karnad, *India's Nuclear Policy*, pp.101–102.

131 B. Karnad, *Why India is not a Great Power*, p.375.

132 This is extrapolated on an estimated production of one to two missiles per year for the Agni-1 and Agni-2 post-developmental trials and a somewhat higher rate for the Agni-III and IV, with Limited Series Production for the Agni-V assumed to have been started. It is probable that each group has 16 launchers and 48 missiles each. This has been gleaned from interviews and from a documentary on Transport Solutions India Limited which noted that a contract for 48 trailers – one for each Agni-V missile intended to be produced – was issued in 2015. See 'Make In India – New Deal For Defence – Transport Solutions India, Episode 9, Segment 1', YouTube video, from 7:20 minutes onwards. https://www.youtube.

com/watch?v=LIaQ3nOGmEI&t=640s (accessible only to Indian users or through a VPN). It is also known that at least 16 Agni-V launchers are on order – see *DRDO Annual Report 2016*, p.39.

133 This is based on two reports – the first in DRDO's annual report of 2016 which indicated that 16 Agni-V launchers had been ordered, and the second a news documentary which noted that 48 trailers (one for each of the Agni-V missiles on order) had been ordered. If one extrapolates that these figures represent a likely average for each Agni missile group, it is therefore probable that India's intended arsenal, in the short term, is likely to be approximately five groups with 48 missiles and 16 launchers each.

134 'Now, India has a nuclear triad', *The Hindu* (18 October 2016), http://www.thehindu.com/news/national/Now-India-has-a-nuclear-triad/article16074127.ece.

135 Ananatha Krishna M.,'K-15 SLBM is a beast with gen-next tech', *New Indian Express* (30 January 2013), http://www.newindianexpress.com/nation/2013/jan/30/k-15-slbm-is-a-beast-with-gen-next-tech-445756.html.

136 V. Pereira, 'N-capable Arihant submarine successfully test-fires unarmed missile', *Magalorean.com* (26 November 2015), http://www.mangalorean.com/n-capable-arihant-submarine-successfully-test-fires-unarmed-missile/.

137 A. Panda, 'India Inches Closer to Credible Nuclear Triad With K-4 SLBM Test', *The Diplomat* (13 May 2014), http://thediplomat.com/2014/05/india-inches-closer-to-credible-nuclear-triad-with-k-4-slbm-test/.

138 Franz-Stefan Gady, 'India Successfully Tests New Ballistic Missile', *The Diplomat* (22 March 2016), http://thediplomat.com/2016/03/india-successfully-tests-new-ballistic-missile/.

139 H.K. Rout, 'EXPRESS EXCLUSIVE: Maiden Test of Undersea K-4 Missile From Arihant Submarine', http://www.newindianexpress.com/nation/2016/apr/09/EXPRESS-EXCLUSIVE-Maiden-Test-of-Undersea-K-4-Missile-From-Arihant-Submarine-921990.html.

140 'India successfully test-fires 3,500 km nuclear-capable missile K-4', *New Indian Express* (20 January 2020), https://www.newindianexpress.com/states/odisha/2020/jan/20/india-successfully-test-fires-3500-km-nuclear-capable-missile-k-4-2091838.html.

141 Rajat Pandit, 'Arihant's N-capable missile "ready to roll"', *Times of India* (25 January 2020), https://timesofindia.indiatimes.com/india/india-successfully-test-fires-k-4-submarine-launched-missile/articleshow/73589861.cms.

142 'India's nuclear triad is complete with INS Arihant ending its first deterrence patrol', *The Hindu* (5 November 2018), https://www.thehindu.com/news/national/ins-arihant-completes-deterrence-patrol-india-declares-nuclear-traid-operational/article25425436.ece.

143 Manu Pubby, 'India's Rs 1.2 lakh crore nuclear submarine project closer to realization', *Economic Times* (21 February 2020), https://economictimes.indiatimes.com/news/defence/indias-rs-1-2-l-cr-n-submarine-project-closer-to-realisation/articleshow/74234776.cms.

144 Sandeep Unnithan, 'From India Today magazine: A peek into India's top secret and costliest defence project, nuclear submarines', *India Today* (18 December 2017), https://www.indiatoday.in/magazine/the-big-story/story/20171218-india-ballistic-missile-submarine-k-6-submarine-launched-drdo-1102085-2017-12-10.

145 'Indian Navy successfully test fires Dhanush missile: All you need to know', *India Today* (26 November 2015), http://indiatoday.intoday.in/education/story/dhanush/1/531950.html.

Chapter 4

1 Badri-Maharaj, *Indian Nuclear Strategy*, pp.156–196.

2 J. Baranwal, *SP's Military Yearbook 1992–93* (New Delhi: Guide Publications, 1993), p.SS13.

3 V. Raghuvanshi, 'Indian Excludes Foreign Vendors From Its Air-Defense Upgrade', *Defense News* (13 June 2016), https://www.defensenews.com/global/asia-pacific/2016/06/13/indian-excludes-foreign-vendors-from-its-air-defense-upgrade/. See also C. Kumar, 'Six new Akash squadrons to give IAF missile muscle', *Times of India* (17 February 2015), https://timesofindia.indiatimes.com/india/Six-new-Akash-squadrons-to-give-IAF-missile-muscle/articleshow/46269673.cms.

4 A. Shukla, 'Army orders surface to air missile, making it the first tri-service weapon', *Business Standard* (26 September 2017), http://www.business-standard.com/article/economy-policy/army-orders-surface-to-air-missile-making-it-the-first-tri-service-weapon-117092500988_1.html.

5 'Defence Acquisition Council okays purchase of Russian air defence missile systems', *Press Trust of India* (17 December 2015), http://www.dnaindia.com/india/report-defence-acquisition-council-okays-purchase-of-russian-air-defence-missile-systems-2156916.

6 P. Sengupta, 'Arudhra MPR Is EL/M-2084 MMR: Seeing Is Believing', *Trishul-Trident* (4 June 2011), http://trishul-trident.blogspot.com/2011/06/arudhra-mpr-is-elm-2084-mmr-seeing-is.html.

7 V. Bhatia, 'IAF to deploy new radar near border areas', *Times of India* (2 October 2011), https://timesofindia.indiatimes.com/city/jaipur/IAF-to-deploy-new-radar-near-border-areas/articleshow/10204373.cms.

8 P. Sengupta, 'IAF's Multi-Phase IACCCS Being Enhanced', *Trishul-Trident* (22 January 2012), http://trishul-trident.blogspot.com/2012/01/iafs-multi-phase-iacccs-being-enhanced.html.

9 Sengupta, 'IAF's Multi-Phase IACCCS Being Enhanced'.

10 Sengupta, 'IAF's Multi-Phase IACCCS Being Enhanced'.

11 'Electronic Weapons: Instant Radar', *Strategy Page* (21 February 2009), https://www.strategypage.com/htmw/htecm/articles/20090221.aspx.

12 P. Sengupta, 'IAF's Multi-Phase IACCCS Being Enhanced', *Trishul-Trident* (22 January 2012), http://trishul-trident.blogspot.com/2012/01/iafs-multi-phase-iacccs-being-enhanced.html.

13 Sengupta, 'IAF's Multi-Phase IACCCS Being Enhanced'.

14 Sengupta, 'IAF's Multi-Phase IACCCS Being Enhanced'.

15 'Electronic Weapons: Instant Radar', n.7

16 Surendra Singh, 'Military using 13 satellites to keep eye on foes', *Times of India* (26 June 2017), https://timesofindia.indiatimes.com/india/military-using-13-satellites-to-keep-eye-on-foes/articleshow/59314610.cms (accessed 3 March 2018).

17 P. Sengupta, 'IAF's Multi-Phase IACCCS Being Enhanced', *Trishul-Trident* (22 January 2012), http://trishul-trident.blogspot.com/2012/01/iafs-multi-phase-iacccs-being-enhanced.html.

18 'India acquires Green Pine radars from Israel', *Press Trust of India* (28 June 2002), https://timesofindia.indiatimes.com/india/India-acquires-Green-Pine-radars-from-Israel/articleshow/14351441.cms.

19 P. Sengupta, 'IAF's Multi-Phase IACCCS Being Enhanced', *Trishul-Trident* (22 January 2012), http://trishul-trident.blogspot.com/2012/01/iafs-multi-phase-iacccs-being-enhanced.html.

20 P. Sawhney, 'Decks Cleared For the Contract Signing of S-400 ADMS in December', *Force India* (17 October 2017), http://forceindia.net/decks-cleared-contract-signing-s-400-adms-december/.

21 'Missile defence shield to be ready in three years: India', *Dawn* (13 December 2007), https://www.dawn.com/news/280120.

22 S. Jha, 'Hit-to-Kill Successfully Demonstrated By DRDO's PDV Interceptor', *Delhi Defence Review* (25 February 2017), http://www.delhidefencereview.com/2017/02/25/hit-kill-successfully-demonstrated-drdos-pdv-interceptor/.

23 T.S. Subramanian, 'Smashing hit"', *Frontline* (22 December 2007 – 4 January 2008), http://www.frontline.in/static/html/fl2425/stories/20080104242512300.htm.

24 'Interceptor missile scores "direct hit"', *The Hindu* (7 December 2007), http://www.hindu.com/2007/12/07/stories/2007120761241800.htm.

25 'Interceptor missile test fired successfully', *NDTV.com* (6 March 2011).

26 'India successfully test-fires AAD missile interceptor', *NDTV.com* (23 November 2012).

27 Y. Mallikarjun, 'Upgraded interceptor missile successfully hits virtual target', *The Hindu* (22 November 2015), http://www.thehindu.com/news/cities/Hyderabad/upgraded-interceptor-missile-successfully-engages-electronically-simulated-target-missile/article7905298.ece.

28 'India successfully test-fires supersonic interceptor missile', *Tribune India* (15 May 2016), http://www.tribuneindia.com/news/nation/india-successfully-test-fires-supersonic-interceptor-missile/237107.html.

29 T.S. Subramanian, 'Interceptor missile mission a "failure"', *The Hindu* (23 May 2016), http://www.thehindu.com/news/national/Interceptor-missile-mission-a-%E2%80%98failure%E2%80%99/article14333898.ece.

30 DDR Staff, 'DRDO's AAD Ballistic Missile Defence Interceptor Heads Toward Induction With Latest Test', *Delhi Defence Review* (29 December 2017), http://www.delhidefencereview.com/2017/12/29/drdos-aad-ballistic-missile-defence-interceptor-heads-toward-induction-with-latest-test/ (accessed 10 January 2018).

31 S. Jha, 'DRDO's AAD Endo-atmospheric Ballistic Missile Interceptor Hits Bullseye', *Delhi Defence Review* (1 March 2017), http://www.delhidefencereview.com/2017/03/01/drdos-aad-endo-atmospheric-ballistic-missile-interceptor-hits-bullseye/.

32 Jha, 'DRDO's AAD Endo-atmospheric Ballistic Missile Interceptor Hits Bullseye'.

33 P. Sengupta, 'AAD Endo-Atmospheric Interceptor Headed For Systems Maturity', *Trishul-Trident Blogspot* (1 January 2018), http://trishul-trident.blogspot.in/2018/01/aad-endo-atmospheric-interceptor-headed.html.

34 Sengupta, 'AAD Endo-Atmospheric Interceptor'.

35 Sengupta, 'AAD Endo-Atmospheric Interceptor'.

36 Air Marshal N. Menon, 'Ballistic Missile Defence System for India', *Indian Defence Review*, Vol. 27.3 (July–September 2012), http://www.indiandefencereview.com/spotlights/ballistic-missile-defence-system-for-india/.

37 'India Successfully Test-Fires New Interceptor Missile', *Outlook India* (27 April 2014), https://web.archive.org/web/20140428163936/http://news.outlookindia.com/items.aspx?artid=838755.

38 M. Somasekhar, 'India's "interceptor" may make ballistic missile shield real', *The Hindu Business Line* (12 February 2017), http://www.thehindubusinessline.com/news/indias-interceptor-may-make-ballistic-missile-shield-real/article9537665.ece.

39 P. Sengupta, 'AAD Endo-Atmospheric Interceptor Headed For Systems Maturity', *Trishul-Trident Blogspot* (1 January 2018), http://trishul-trident.blogspot.in/2018/01/aad-endo-atmospheric-interceptor-headed.html.

40 Snehesh Alex Philip, 'India's ballistic missile shield ready, IAF & DRDO to seek govt nod to protect Delhi', *The Print* (8 January 2020), https://theprint.in/defence/indias-ballistic-missile-shield-ready-iaf-drdo-to-seek-govt-nod-to-protect-delhi/345853/.

Chapter 5

1 Badri-Maharaj, *Indian Nuclear Strategy*, pp.279–286.

ABOUT THE AUTHOR

Sanjay Badri-Maharaj, from Trinidad and Tobago, received his MA and PhD from the Department of War Studies, Kings College London. His thesis was on India's Nuclear Weapons Program. He has written and published extensively, including *The Armageddon Factor: Nuclear Weapons in the India-Pakistan Context* (2000) and *Indian Nuclear Strategy: Confronting the Potential Nuclear Threat from both Pakistan and China* (2018), as well as *Kargil 1999*, *Trinidad 1990* and *Operation Meghdoot* for Helion's @War series. He has served as a consultant to the Ministry of National Security in Trinidad and was a visiting International Fellow at the Institute for Defence Studies and Analyses, New Delhi.